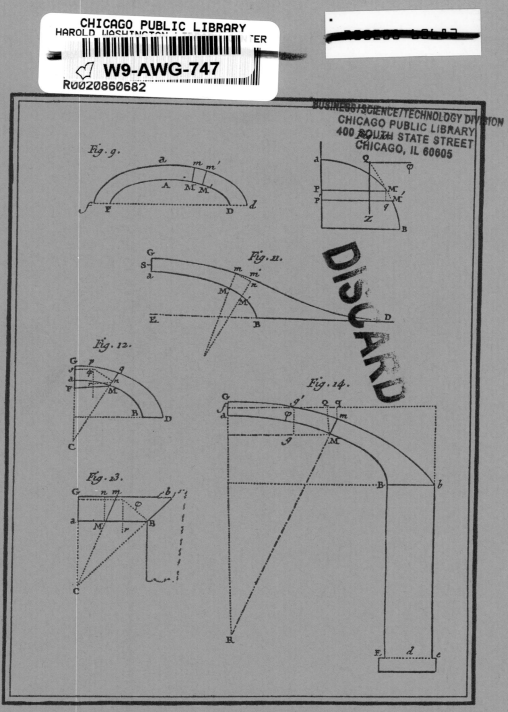

Fig. 9.

Fig. 11.

DISCARD

Fig. 12.

Fig. 14.

Fig. 13.

COULOMB'S
MEMOIR ON STATICS

Portrait of Coulomb

Photograph: Conservatoire National des Arts et Métiers

COULOMB'S MEMOIR ON STATICS

*An Essay in the History
of Civil Engineering*

JACQUES HEYMAN

Professor of Engineering, University of Cambridge

*Cambridge
at the University Press 1972*

Published by the Syndics of the Cambridge University Press
Bentley House, 200 Euston Road, London NW1 2DB
American Branch: 32 East 57th Street, New York, N.Y.10022

© Cambridge University Press 1972

Library of Congress Catalogue Card Number: 72-75771

ISBN: 0 521 08395 8

Printed in Great Britain
at the University Printing House, Cambridge
(Brooke Crutchley, University Printer)

Contents

Preface		*page vii*
1	*The ESSAI*	*1; 41*
	Notes	*70*
2	*Coulomb's References*	*75*
	Gregory	*75*
	Amontons	*76*
	Bossut	*77*
	Musschenbroek	*80*
	Euler	*81*
	James Bernoulli	*81*
	La Hire	*82*
	Bélidor	*84*
3	*The Strength and Stiffness of Beams*	*89*
	The strength of beams (1)	*90*
	Coulomb's *Essai*	*97*
	The strength of beams (2)	*100*
	Navier and Saint-Venant	*101*
	The stiffness of beams	*105*
4	*Coulomb's Equation*	*110*
	The Mohr–Coulomb criterion	*113*
	Upper bounds from velocity fields	*116*
	Coulomb's material tests	*120*
5	*The Thrust of Soil*	*124*
	Coulomb's problem (1); $c = 0$	*132*
	Coulomb's problem (2); (c, ϕ)	*137*
	The stability of a vertical cut (1); $\phi = 0$	*140*
	The stability of a vertical cut (2); (c, ϕ)	*143*

v

Thrust against a rough retaining wall; $c = 0$ *page 144*
Drucker's theorems   *147*
Nineteenth-century work   *148*
Sokolovskii   *156*
Critical state theory   *160*

6   *The Thrust of Arches*   *162*

The plastic theorems applied to masonry   *162*
Eighteenth-century work (1)   *168*
Couplet   *170*
Eighteenth-century work (2)   *175*
Coulomb and Bossut   *181*
The nineteenth century and later   *183*
Scholium   *189*

7   *Some Historical Notes*   *190*

Technical education in France   *191*
Coulomb's early life   *194*
The *Essai*   *195*
Coulomb's later life   *197*

References   *199*

Name Index   *207*

Subject Index   *210*

Preface

Coulomb's *Essai* on 'some statical problems' is most widely known as the memoir which laid the foundations of the modern science of soil mechanics; the journal *Géotechnique*, for example, uses the cul-de-lampe on p. 40 of the *Essai* on the cover of each of its issues. However, Coulomb discusses in the same paper three other major problems of eighteenth-century civil engineering, namely the bending of beams, the fracture of columns, and the calculation of abutment thrusts developed by masonry arches. Knowing of my interest in the last topic, Dr A. N. Schofield suggested that we should prepare a critical edition of the *Essai*, with notes on each of the four problems. However, he was appointed almost immediately to a Professorship at Manchester, and the demands of a large and active department have prevented the collaboration that we both wished.

Professor Schofield has been good enough to read an early draft of the whole book, and final drafts of chapters 4 and 5, but he cannot be held in any way responsible either for the present form of the book or for any particular statements. Indeed, I am conscious that chapter 5, which is the mathematical history of a single problem in soil mechanics (that of the thrust of soil against a retaining wall), would have had a very different content had he written the first draft, as we originally intended. In its present form, the chapter notes the first application of the principles of mechanics to the evaluation of earth pressures (by Bullet in 1691), and then follows the thread of this work through the eighteenth century up to Coulomb. The story is continued into the nineteenth century until the theory has reached a form (with Boussinesq in 1882) which is recognisably modern. There is a brief discussion of the work of Sokolovskii in the present century, which represents the final development of this type of theory.

In the whole of chapter 5 soil is assumed to be a single-phase material, having constant cohesion c and angle of internal friction ϕ; this was in fact Coulomb's own assumption. As far as possible, the treatment is brought within the framework of the limit theorems of the theory of plasticity, and it is really more true to say that, instead of the thread being followed forward from 1691, the antecedents of the modern single-phase theory are traced back to Coulomb and beyond.

Ever since Terzaghi introduced the idea of soil as a two-phase material, it has become apparent that the single-phase treatment is a poor model of many problems in soil mechanics; Schofield's name is associated with the recent development of Critical State Theory, which takes into account not only cohesion and friction, but also the water content of the soil. However, the historical examination of the antecedents of this theory is not rewarding; although engineers were aware of the importance of water, there is very little to be found in the literature beyond occasional warnings (e.g. by Bossut in 1762, or by Coulomb himself, p. 23; p. 57).*

The form of chapter 5 is roughly repeated in chapters 3 and 6; the theory of the bending of beams is followed from Galileo through Coulomb to Saint-Venant, and that of the thrust of arches from La Hire and Couplet to the modern application of the limit theorems of plasticity. Chapter 4 is almost entirely non-historical, and is really an introduction to the way the ideas of plasticity can be used to discuss Coulomb's problems in general, and the fracture of columns in particular.

Although some of the older work appears to be newly rediscovered here, this whole book leans very heavily on the findings of previous commentators. Interesting early historical accounts are given by Poleni, Girard, and Mayniel for the problems of arches, bending of beams, and soil mechanics respectively; in the nineteenth century, both Augoyat and Maindron are helpful in the understanding of French technical education, a subject which has been treated recently by Artz. The more recent technical histories are quoted in the text and listed in the references; of these, Truesdell's account of flexible bodies is definitive.

It is of interest to give some account of Coulomb's life, and it is most fortunate that Gillmor's book is now available. Previous accounts were superficial, and the historical notes included in chapter 7, while very brief, are at least based on direct archival research.

* Here and throughout the book, the first page number refers to the French text, and the second to the English translation.

COULOMB, C. A., Essai sur une application des règles *de maximis & minimis* à quelques problèmes de statique, relatifs à l'architecture, *Mémoires de Mathématique & de Physique, présentés à l'Académie Royale des Sciences par divers Savans, & lûs* dans ses Assemblées*, vol. 7, 1773, pp. 343–82, Paris (1776).

Reprinted in *Théorie des machines simples*, Paris (1821).

* Coulomb's *Mémoire* was read by him on 10 March and 2 April 1773.

E S S A I

Sur une application des règles de Maximis & Minimis à quelques Problèmes de Statique, relatifs à l'Architecture.

Par M. COULOMB, Ingénieur du Roi.

INTRODUCTION.

CE Mémoire est destiné à déterminer, autant que le mélange du Calcul & de la Physique peuvent le permettre, l'influence du frottement & de la cohésion, dans quelques problèmes de Statique. Voici une légère analyse des différens objets qu'il contient.

Après quelques observations préliminaires sur la cohésion, & quelques expériences sur le même objet, l'on détermine la force d'un pilier de maçonnerie; le poids qu'il peut porter, pressé suivant sa longueur; l'angle sous lequel il doit se rompre. Comme ce problème n'exige que des considérations assez simples, qui servent à faire entendre toutes les autres parties de cet Essai, tâchons de développer les principes de sa solution.

Si l'on suppose un pilier de maçonnerie coupé par un plan incliné à l'horizon, en sorte que les deux parties de ce pilier soient unies dans cette section, par une cohésion donnée, tandis que tout le reste de la masse est parfaitement solide, ou lié par une adhérence infinie; qu'ensuite on charge ce pilier d'un poids: ce poids tendra à faire couler la partie supérieure du pilier sur le plan incliné, par lequel il touche la partie inférieure. Ainsi, dans le cas d'équilibre, la portion de la pesanteur, qui agit parallèlement à la section, sera exactement égale à la cohérence. Si l'on remarque actuellement, dans le cas de l'homogénéité, que l'adhérence du pilier est réellement égale pour toutes les parties; il faut, pour que le pilier puisse supporter un fardeau, qu'il n'y ait aucune section de ce pilier, sur laquelle l'effort décomposé de sa pression puisse faire couler la partie supérieure. Ainsi, pour déterminer le plus grand poids que puisse supporter un pilier, il faut chercher parmi toutes ses sections celle dont la cohésion est en équilibre avec un poids qui soit un *minimum :* car, pour lors, toute pression, au-dessus de celle déterminée par cette condition, seroit insuffisante pour rompre le pilier.

Outre la résistance qui provient de la cohésion, j'ai eu égard à celle dûe au frottement. Les mêmes principes suffisent pour remplir les deux conditions: l'application de cette recherche peut s'étendre à tous nos édifices, dont la masse est toujours soutenue par des colonnes, ou par quelque moyen équivalent.

L'on détermine ensuite la pression des terres, contre les plans verticaux qui les soutiennent; la méthode est absolument la même. Si l'on suppose en effet un triangle-rectangle solide, dont un des côtés, soit vertical, & dont l'hypothénuse touche un plan incliné, sur lequel le triangle tend à glisser; si ce triangle, sollicité par sa pesanteur, est soutenu par une force horizontale, par sa cohésion, & par son frottement, qui agissent le long de cette hypothénuse, l'on déterminera facilement, dans le cas d'équilibre, cette force horizontale par les principes de Statique. Si l'on remarque ensuite que les terres étant supposées homogènes, peuvent se séparer dans le cas de rupture, non-seulement suivant une ligne droite, mais suivant une ligne courbe quelconque; il s'ensuit que pour avoir la pression d'une surface de terre contre un plan vertical, il faut trouver parmi toutes les surfaces décrites dans un plan indéfini vertical, celle qui, sollicitée par sa pesanteur, & retenue par son frottement & sa cohésion, exigeroit, pour son équilibre, d'être soutenue par une force horizontale, qui fut un *maximum;* car, pour lors il est évident que toute autre figure demandant une moindre force horizontale, dans le cas d'équilibre, la masse adhérente ne pourroit

se

se divifer. Comme l'expérience donne à peu-près une ligne droite pour la ligne de rupture des terres, lorfqu'elles ébranlent leurs revêtemens, il fuffit, dans la pratique, de chercher dans une furface indéfinie, parmi tous les triangles qui preffent un plan vertical, celui qui demande, pour être foutenu, la plus grande force horizontale. Dès que cette force eft déterminée l'on en déduit avec facilité les dimenfions des revêtemens.

L'on trouvera à la fin de ce même article les moyens de déterminer exactement parmi toutes les furfaces courbes que l'on peut tracer dans un fluide indéfini, celle dont la preffion contre un plan vertical, eft un *maximum,* en ayant égard au frottement & à la cohéfion. Cette recherche peut fervir à trouver la preffion des fluides cohérens, contre les parois des vafes qui les foutiennent.

Enfin on termine ce Mémoire par chercher les dimenfions des voûtes, leurs points de rupture, les limites qui circonf-crivent leur état de repos, lorfque la cohéfion & le frottement contribuent à leur folidité. M. Gregori a démontré, je crois le premier, dans les *Tranfactions Philofophiques,* que dans le fyftème de la pefanteur, la chaînette étoit la même courbe que la voûte qui feroit formée par une infinité d'élémens d'une épaiffeur conftante & infiniment petite. J'ai étendu cette propofition, & j'ai prouvé que, quel que fût le nombre & la direction des forces qui agiroient fur une voûte formée d'après les fuppofitions précédentes, la figure de cette voûte feroit la même que celle d'une chaînette follicitée par les mêmes puiffances. Les mêmes principes fuffifent enfuite pour déterminer les joints lorfqu'ils font des quantités finies, ou qu'ils doivent former avec la courbe intérieure de la voûte un autre angle que le droit. Cette dernière hypothèfe a lieu dans les plates-bandes; l'on y trouve que fi l'épaiffeur eft donnée, les joints, dans le cas d'équilibre, doivent être dirigés vers un même centre.

Les formules trouvées, en faifant abftraction des frottemens & de la cohéfion des joints, ne peuvent être d'aucune utilité dans la pratique; tous les Géomètres qui se sont occupés de cet objet s'en sont aperçus; ainsi, pour avoir des résultats que l'on peut employer, ils ont été obligés de fonder leurs calculs sur des suppositions qui les rapprochassent de la Nature. Ces suppositions consistent ordinairement à considérer les voûtes comme divisées en plusieurs parties, & à chercher ensuite les conditions d'équilibre de ces différentes parties : mais comme cette division se fait à peu-près, d'une manière arbitraire; dans le dessein de l'apprécier, j'ai cherché par les règles *de maximis & minimis*, quels seroient les véritables points de rupture dans les voûtes trop foibles, & les limites des forces que l'on pourroit appliquer à celle dont les dimensions seroient données : j'ai tâché autant qu'il m'a été possible de rendre les principes dont je me suis servi assez clairs pour qu'un Artiste un peu instruit pût les entendre & s'en servir.

Ce Mémoire, composé depuis quelques années, n'étoit d'abord destiné qu'à mon usage particulier, dans les différens travaux dont je suis chargé par mon état; si j'ose le présenter à cette Académie, c'est qu'elle accueille toujours avec bonté le plus foible essai, lorsqu'il a l'utilité pour objet. D'ailleurs, les Sciences sont des monumens consacrés au bien public; chaque citoyen leur doit un tribut proportionné à ses talens. Tandis que les grands hommes, portés au sommet de l'édifice, tracent & élèvent les étages supérieurs, les artistes ordinaires répandus dans les étages inférieurs, ou cachés dans l'obscurité des fondemens, doivent seulement chercher à perfectionner ce que des mains plus habiles ont créé.

PROPOSITIONS PRÉLIMINAIRES.

I.

Fig. 4.   Soit le plan *a b c d e*, sollicité par des forces quelconques situées dans la direction de ce plan, en équilibre sur la ligne AB; la résultante de toutes ces forces sera perpendiculaire à la ligne AB, & tombera entre les points *a* & *e*.

I I.

Si toutes les forces, qui agiſſent dans ce plan ſont décompoſées ſuivant deux directions, l'une parallèle à AB, l'autre
qui lui ſoit perpendiculaire, la ſomme des forces décompoſées,
parallèlement à AB, ſera nulle, & la ſomme des forces,
perpendiculaires à AB, égalera la preſſion qu'éprouve la
ligne AB.

I I I.

Si la preſſion qu'éprouve la ligne AB eſt exprimée
par P, le même plan pourra être ſuppoſé ſollicité par toutes
les forces qui lui ſont appliquées, & de plus par la réaction
de la preſſion. Mais ſi toutes ces forces, ainſi que la réaction
de la préſſion, ſont décompoſées ſuivant deux directions
quelconques perpendiculaires l'une à l'autre ; il ſuit de l'équilibre & de la perpendicularité des deux directions, que la
réſultante ſuivant chaque direction, ſera nulle.

I V.

Du Frottement.

Le frottement & la cohéſion ne ſont point des forces
actives comme la gravité, qui exerce toujours ſon effet en
entier, mais ſeulement des forces coërcitives ; l'on eſtime
ces deux forces par les limites de leur réſiſtance. Lorſqu'on
dit, par exemple, que dans certains bois polis, le frottement
ſur un plan horizontal d'un corps peſant neuf livres, eſt
trois livres ; c'eſt dire que toute force au-deſſous de trois
livres ne troublera point ſon état de repos.

Je ſuppoſerai ici que la réſiſtance dûe au frottement eſt
proportionnelle à la preſſion, comme l'a trouvé M. Amontons ;
quoique dans les groſſes maſſes le frottement ne ſuive pas
exactement cette loi. D'après cette ſuppoſition, l'on trouve
dans les briques le frottement, les trois quarts de la preſſion.
Il ſera bon de faire des épreuves ſur les matériaux que l'on
voudra employer. Il eſt impoſſible de fixer ici le frottement

X x ij

des pierres, les essais faits pour une carrière ne pouvant point servir pour une autre.

V.

De la cohésion.

La cohésion se mesure par la résistance que les corps solides opposent à la désunion directe de leurs parties. Comme chaque élément des solides, lorsqu'ils sont homogènes, est doué de cette même résistance; la cohésion totale est proportionnelle au nombre des parties à désunir, & par conséquent a la surface de rupture des corps. J'ai cherché à déterminer par quelques expériences, la force de cette cohésion; elles m'ont donné les résultats suivans.

I.ere Expérience. Fig. 1.

J'ai pris un carreau *a b c d*, d'une pierre blanche, d'un grain fin & homogène *; ce carreau étoit d'un pied quarré, avoit un pouce d'épaisseur; je l'ai fait échancrer en *e* & en *f*, en sorte que *ef* formoit une gorge de deux pouces, par laquelle les deux parties du carreau restoient unies. J'ai suspendu ce carreau par cette gorge, en y introduisant deux cordes nouées en fronde; & par deux autres cordes j'ai suspendu un plateau de balance que j'ai chargé d'un poids *P*. Il a fallu augmenter ce poids jusqu'à 430 livres, pour rompre le carreau en *ef*, ce qui donne, pour la force de la cohésion, 215 livres par pouces.

II.eme Expérience.

J'ai voulu voir si en rompant un solide de pierre, par une force dirigée suivant le plan de rupture, il falloit employer le même poids que pour le rompre, comme dans l'expérience précédente, par un effort perpendiculaire à ce

Fig. 2.

plan. Pour cela j'ai introduit le petit solide *ABCD* dans une mortoise *AGeg*, j'ai suspendu un bassin à la corde *eP*, qui enveloppoit le solide & qui joignoit la mortoise; le petit solide avoit deux pouces de largeur, un pouce de hauteur, ce qui donne la même surface de rupture que dans l'expérience

---

* Cette pierre se trouve autour de Bordeaux, & sert à construire les façades des grands édifices de cette ville.

précédente; il n'a rompu que lorfque le baffin a été chargé de 440 livres. J'ai répété plufieurs fois cette expérience, de même que la première, & j'ai prefque toujours trouvé qu'il falloit une plus grande force pour rompre le folide, lorfque cette force étoit dirigée fuivant le plan de rupture, que lorfqu'elle étoit perpendiculaire à ce plan. Cependant, comme cette différence n'eft ici que $\frac{1}{44}$ du poids total, & qu'elle s'eft trouvée fouvent plus petite, je l'ai négligée dans la théorie qui fuit.

J'ai voulu voir comment fe fait la rupture d'un corps, lorfqu'il eft rompu par une force qui agit fur lui avec un bras de levier; en conféquence, j'ai encaftré dans une mortoife $ACeg$ un folide de la même pierre que dans l'expérience précédente, ayant 1 pouce de hauteur, 2 pouces de largeur, & 9 pouces de longueur de g en D, où j'ai fufpendu un poids P; ce poids s'eft trouvé de 20 livres lorfque le folide a caffé en eg.

III.ᵉ Expérience.

Fig. 3.

V I.

J'ai répété les mêmes épreuves fur des briques de Provence d'une excellente cuite & d'un grain très-uni, j'ai trouvé que leur cohéfion, en les rompant par une force perpendiculaire au plan de rupture, conformément à la première expérience, étoit de 280 à 300 livres par pouces. J'ai trouvé encore qu'un mortier compofé de quatre parties de fable & trois de chaux, employé depuis deux ans, fupportoit, perpendiculairement au plan de rupture, 50 livres par pouces. Cette dernière épreuve, faite à la Martinique ne peut point être généralifée; la force du mortier varie quelquefois du double, & même du triple, fuivant la nature du pays humide ou fec, fuivant les qualités du fable, de la chaux, de la pierre employée dans le corps de la maçonnerie, fuivant l'ancienneté de cette maçonnerie; l'on ne peut rien fixer, il faut dans chaque lieu des obfervations particulières.

VII.

Remarques fur la rupture des Corps.

Fig. 6.   Si l'on fuppofe un folide *o n KL* dont les angles foient droits, alongé comme une poutre ordinaire, & fixé en *o n*, de manière que les côtés de ce folide foient horizontaux & verticaux ; fi l'on fuppofe enfuite que ce folide eft coupé par un plan vertical repréfenté par *AD*, perpendiculaire au côté *o n KL*, & follicité par un poids φ, attaché à fon extrémité en *L*; il eft évident, en ne confidérant qu'une face verticale de ce folide, les autres étant égales & parallèles, que tous les points de la ligne *AD* réfiftent pour empêcher le poids φ de rompre le folide; que par conféquent une partie fupérieure *AC* de cette ligne fait effort par une traction dirigée fuivant *QP*, tandis que la partie inférieure fait effort, par une preffion dirigée fuivant *Q'P'*. Si l'on décompofe toutes les forces, foit de traction, foit de preffion, fuivant deux directions, l'une verticale & l'autre horizontale, exprimée par *QM* & *PM;* & fi par tous les points *M* l'on fait paffer une ligne *BMCe*, cette courbe fera le lieu géométrique de tous les efforts perpendiculaires qu'éprouve la ligne *AD*. Ainfi, la tranche *ADKL* doit être fuppofée follicitée par toutes les forces horizontales *PM*, par toutes les forces verticales *MQ*, & par la pefanteur du poids φ; par conféquent, puifqu'il y a équilibre, il faut, *art. 3*, que la fomme des puiffances horizontales foit nulle; que, par conféquent, l'aire des tenfions *ABC* égale l'aire des preffions *Ced*. Il faut de plus, par le même article, que la fomme des forces verticales *QM* foit égale au poids φ; mais par les principes de Statique l'on a encore la fomme des *momentum* autour du point *G* de toutes les forces, foit de traction, foit de preffion, égale au *momentum* du poids φ autour du même point; ce qui donne l'équation $\int Pp \cdot MP \cdot CP = \varphi LD$. Nous avons donc, quel que foit le rapport entre la dilatation des élémens d'un folide & leur cohéfion, les trois conditions précédentes à remplir.

Je fuppofe, par exemple, que l'on veuille chercher le poids que peut fupporter une pièce de bois parfaitement élaftique; c'eft-à-dire qui fe comprime ou fe dilate chargée dans la direction de fa longueur, proportionnellement à la force qui la comprime ou qui la dilate; que l'élément $ofnh$, qui touche le mur, repréfente une portion très-petite de la pièce de bois dans fon état naturel; fi l'on charge cette pièce de bois d'un poids φ, la partie fupérieure de la ligne fh fe portera en g, & la partie inférieure fe portera en m; la ligne fh deviendra gm: mais comme, par hypothèfe, les tenfions, de même que les preffions, font repréfentées par les parties $\pi\mu$ du triangle fge, il fuit que le triangle de compreffion emh doit égaler le triangle de dilatation fge. Ainfi, fi l'on nomme δ la tenfion du point f, repréfentée par fg, fe égalera $\frac{1}{2}fh$; l'on aura, pour le *momentum* du petit triangle de traction, $\frac{\delta ef^2}{3}$, qui, ajouté au *momentum* du petit triangle de compreffion, doit donner $\frac{\delta(fh)^2}{6} = \varphi n.L$, ou δfh, dans l'inftant de rupture, exprime la réfiftance que l'adhérence oppoferoit à un effort qui agiroit perpendiculairement à la ligne fh, en fuppofant cependant que les tractions MQ n'influent que très-peu fur la réfiftance des folides; ce qui eft affez vrai, lorfque le bras de levier nL du poids φ eft beaucoup plus grand que l'épaiffeur fh.

Mais fi l'on fuppofoit le folide, prêt à fe rompre, compofé de fibres roides, ou qui ne foient fufceptibles ni de compreffion, ni d'alongement; fi l'on fuppofoit encore que le corps fe rompit en tournant autour du point h; pour lors, chaque point de l'épaiffeur fh feroit un effort égal; le point h éprouveroit une preffion égale à δfh, & le *momentum* du petit triangle de cohéfion feroit $\frac{\delta(fh)^2}{2}$. Appliquons cette dernière hypothèfe à nos expériences.

J'ai trouvé par la première expérience, qu'une furface de

Fig. 2. deux pouces de largeur fur un pouce de hauteur, oppofoit une réfiftance égale à 430 livres. Dans la troifième expérience j'ai les mêmes dimenfions, & de plus hL égale 9 pouces; par conféquent, fi la dernière hypothèfe étoit vraie, j'aurois dû trouver $P = \dfrac{430}{2.9}$, à peu-près 24 livres; mais l'expérience donne pour P, 20 livres; ainfi l'on ne peut pas fuppofer dans la rupture des pierres, ou que la roideur des fibres foit parfaite, ou que le point d'appui de rotation foit précifément en h. Une remarque affez fimple auroit fait prévoir ce réfultat, c'eft qu'en prenant h pour point de rotation, il faudroit que ce point h fupportât une preffion finie, fans que fa cohéfion fut détruite, ce qui n'eft pas poffible, puifque cette cohéfion eft une quantité finie, pour une furface finie. Il faut donc, dans le cas qui précède celui de rupture, que cette force, porte en un point h', tel que l'adhérence de $h'q$, foit en état de fupporter par fa réfiftance la preffion $\int f h'$, qu'éprouve la ligne hh', décompofée fuivant $h'q$. Nous donnerons dans la fuite les moyens de déterminer l'angle q du triangle $h'hq$.

M. l'Abbé Boffut, dans un excellent Mémoire fur la figure des digues, ouvrage où l'on trouve réunie, à l'efprit d'invention, la fagacité du Phyficien, & l'exactitude du Géomètre, paroît avoir diftingué & fixé le premier la différence qui fe trouve entre la rupture des bois & celle des pierres.

VIII.

Réfiftance des Piliers de Maçonnerie.

Soit un pilier homogène de maçonnerie, que je fuppofe Fig. 5. d'abord quarré, chargé d'un poids P; l'on demande la direction de la ligne CM, fuivant laquelle ce pilier fe rompra, & la pefanteur du poids néceffaire pour cette rupture.

Je fuppofe ici que l'adhérence oppofe une égale réfiftance, foit que la force foit dirigée parallèlement ou perpendiculairement au plan de rupture, conformément à la première & deuxième expérience. Je fuppofe encore le pilier d'une matière

matière homogène, dont la cohésion soit δ; soit prise une
section quelconque CM, inclinée à l'horizon, & perpendiculaire au plan vertical $ABDM$, face de ce pilier. Si l'on
suppose pour un instant l'adhérence de la partie supérieure
$ABCM$ infinie, de même que celle de la partie inférieure CDM, il est clair que la masse de cette colonne
tendroit à glisser le long de CM; & par conséquent, si les
deux parties étoient unies par une force d'adhérence égale à
la cohésion naturelle du pilier, pour rompre cette colonne,
suivant CM, il faudroit que la pesanteur du poids P, décom-
posée suivant cette direction, fût égale, ou plus grande que
l'adhérence de CM. Soit l'angle en $M\ldots x$, $DM\ldots a$,
P le poids dont la pression représentée par φq, se décompose
suivant les directions φr & rq, perpendiculaires & parallèles
à la ligne de rupture. Si l'on néglige, pour simplifier, la
pesanteur de la colonne, l'on aura $\delta CM = \dfrac{\delta a}{\text{cos. } x}$,
& $rq = P$ sin. x; par conséquent, dans le cas d'équilibre,
l'on trouve $P = \dfrac{\delta a a}{\text{sin. } x . \text{cos. } x}$; mais comme la colonne doit
être en état de porter le poids P sans se rompre, quelle que
soit la section CM, il faut que le poids P soit toujours plus
petit que la quantité $\dfrac{\delta a a}{\text{sin. } x \cos. x}$, quelle que soit la valeur
de x; ce qui aura lieu lorsque l'on déterminera P, tel qu'il
soit un *minimum*, d'après l'équation $P = \dfrac{\delta a^2}{\text{sin. } x . \text{cos. } x}$; ce
qui donne $dP = \dfrac{\delta a a \left[- dx (\text{cos. } x)^2 + dx (\text{sin. } x)^2 \right]}{(\text{sin. } x . \text{cos. } x)^2}$, & par
conséquent sin. $x = $ cos. x. Ainsi le plus grand poids que la
colonne puisse supporter sans se rompre, égale $2 \delta a a$,
le double de la résistance qu'elle opposeroit à une force de
traction, & l'angle de moindre résistance, ou de rupture,
sera 45 degrés.

Nous avons supposé dans cette recherche, que la section
représentée par CM étoit perpendiculaire au côté vertical

ABDM; mais l'on auroit trouvé les mêmes réfultats pour une fection quelconque, pourvu qu'elle eût eu la même inclinaifon fur le plan horizontal; en remarquant que par la théorie des projections, les fections obliques d'un pilier font à leur projection horizontale comme le rayon eft au cofinus d'inclinaifon de ces deux plans; ainfi, en nommant x le finus d'inclinaifon de ces deux plans, & A la furface de la bafe, égale ici à a^2, l'on aura, pour l'adhérence de la fection oblique $\frac{\delta a a}{\text{cof.} x}$, & P fin. x, pour la force qui tend à faire couler la partie fupérieure de la colonne fur le plan incliné qui lui fert de bafe, de quelque manière que foit fitué le plan de fection. Comme ces quantités font précifément les mêmes que les précédentes, elles doivent, par conféquent, donner les mêmes réfultats; d'où l'on peut conclure que, quelle que foit la figure de la bafe horizontale d'un pilier, fi la furface de cette bafe eft conftante, fa force fera la même.

I X.

Nous n'avons point fait entrer, dans la folution précédentes, le frottement qui s'oppofe à la rupture du pilier. Si l'on vouloit y avoir égard, en confervant les dénominations précédentes, l'on trouveroit, pour la preffion du poids fur CM, P cof. x; & comme le frottement eft proportionnel à la preffion, il fera égal à $\frac{P \text{cof.} x}{n}$, n étant une quantité conftante; la maffe du pilier $ABCM$, preffé par le poids P, eft donc retenue par la cohéfion & par le frottement; ainfi, en augmentant le poids jufqu'à ce qu'il foit prêt à rompre le pilier, l'on aura $\frac{a a \delta}{\text{cof.} x} + \frac{P \text{cof.} x}{n} = P$ fin. x, & $P = \delta a a :$ $[\text{cof.} x \left(\text{fin.} x - \frac{\text{cof.} x}{n} \right)]$. Il faut, par les principes qui précèdent, pour avoir le poids que le pilier peut porter fans fe rompre, faire P un *minimum*, ce qui donne

$$dx \left[\text{fin.}\, x \left(\text{fin.}\, x - \frac{\text{cof.}\, x}{n} \right) \right] - dx \, \text{cof.}\, x \left(\text{cof.}\, x + \frac{\text{fin.}\, x}{n} \right) = 0,$$

& par conféquent $(\text{cof.}\, x)^2 + \dfrac{2\,\text{fin.}\, x\,\text{cof.}\, x}{n} = (\text{fin.}\, x)^2$;

d'où l'on tire $\text{cof.}\, x = \text{fin.}\, x \left[\sqrt{\left(1 + \frac{1}{nn} \right)} - \frac{1}{n} \right]$;

d'où $\text{tang.}\, x = \dfrac{1}{\sqrt{\left(1 + \frac{1}{nn} \right)} - \frac{1}{n}}$.

Si le pilier étoit de brique, l'on auroit *(art. 4)* $\frac{1}{n}$ $= \frac{3}{4}$. $\text{tang.}\, x = 2$, $\text{fin.}\, x = 2\,\text{cof.}\, x$; par conféquent, $\text{cof.}\, x = \left(\frac{1}{5} \right)^{\frac{1}{2}}$, & $P = \dfrac{\delta\, a a}{\text{cof.}\, x \,(2\, \text{cof.}\, x - \frac{1}{2}\,\text{cof.}\, x)} = 4\,\delta\, a a$, l'angle en *M* fera de $63^{\text{d}}\, 26'$; ainfi, la force qu'il faudroit pour rompre une colonne de brique par une force preffante, feroit quadruple de celle qu'il faudroit pour rompre cette même colonne par une force de traction.

M. Muffchenbroëk *(Effai de Phyfique, traduction françoife, vol. I, page 354)* a trouvé qu'un pilier quarré de brique, de 11 pouces & demi de longueur fur 5 lignes de côté, a été rompu par un fardeau de 195 livres. Dans l'expérience de M. Muffchenbroëk, les côtés étant $\frac{5}{12}$ de pouce, la coupe horizontale étoit $\frac{25}{144}$ d'un pouce quarré. Or, par l'*art. 6*, nous avons trouvé qu'un pouce quarré de brique fupporte, perpendiculairement au plan de rupture, 300 livres ; ainfi, dans cette expérience $\delta\, a a = 300^{\text{l}} : \frac{25}{144} = 52^{\text{l}}$, qui exprime la force de traction ; mais comme $P = 4\,\delta\, a^2$, il fuit de notre théorie & de nos épreuves, que ce Phy-ficien auroit dû trouver 208 livres, quantité peu différente de 195 livres, réfultat de fon expérience.

Au refte, je fuis obligé d'avertir que la manière dont M. Muffchenbroëk détermine la force d'un pilier de maçon-nerie, n'a aucun rapport avec celle que je viens d'employer. Un pilier, preffé par une force dirigée fuivant fa longueur, ne fe rompt, dit ce Phyficien célèbre, que parce qu'il commence à fe courber ; autrement il fupporteroit toute

Y y ij

forte de poids. En partant de ce principe, il détermine la force des piliers quarrés, en raison inverse du quarré de leur longueur, & triplée de leurs côtés; en sorte que si le pilier dont nous venons de calculer la force n'avoit eu que la moitié de sa première longueur, il auroit supporté un poids quadruple du premier, c'est-à-dire, 832 livres; au lieu que je crois avoir démontré qu'il n'auroit guère supporté que le même poids de 208 livres.

L'on conclud de la formule, que les forces des piliers homogènes sont entr'elles comme les sections horizontales.

L'on détermineroit, par les mêmes principes, l'angle de rupture d'une colonne incompressible, qui seroit pressée par une force inclinée à sa base horizontale; pourvu que la direction de cette force tombât dans cette base; car si elle tomboit en dehors de cette base, il y auroit quelques autres considérations qui rendroient la solution de ce Problème un peu plus difficile.

L'on trouve aussi, par les principes précédens, la hauteur où l'on peut élever une tour sans qu'elle s'écrase sous son propre poids. Supposons, pour simplifier, que cette hauteur est beaucoup plus grande que la largeur; pour pouvoir négliger le petit prisme *CDM,* il faudra substituer dans les formules, à la place de la quantité *P,* la masse d'une tour qui auroit le même poids: supposons-là, par exemple, construite en briques; le pied cube de brique pesant à peu-près 144 livres, un petit prisme, qui auroit un pouce de base, sur un pied de hauteur, pèseroit une livre; ainsi, comme une base d'un pouce peut supporter une force de traction égale à 300 livres, & une force de pression double, lorsque l'on néglige le frottement, il est clair qu'en substituant à la tour une masse de petits prismes, d'un pouce de base, sur 600 pieds de hauteur, il seroit aussi soutenu par la cohérence Si l'on avoit égard au frottement, l'on pourroit, par les mêmes principes, élever cette tour jusqu'à 1200 pieds de hauteur: si à la place de la tour on substituoit une pyramide, elle pourroit s'élever à une hauteur triple.

Si cette tour étoit portée fur plufieurs piliers, la hauteur à laquelle on pourroit l'élever, feroit en raifon directe de la fection horizontale de ces piliers ; en forte que fi la fection de ces piliers étoit, par exemple, le fixième de la fection horizontale de la tour, elle ne pourroit s'élever au-deffus des colonnes qu'à 100 pieds de hauteur, en négligeant le frottement, & à 200 pieds en y ayant égard. L'on néglige ici le poids des piliers, il feroit facile d'y avoir égard.

Lorfque plufieurs voûtes prennent leur naiffance fur le même pilier, s'arc-boutent & fe foutiennent mutuellement, quant à la preffion horizontale ; la réfultante de leurs forces étant verticale, & dirigée fuivant l'axe du pilier, l'on déterminera facilement par cette méthode, la groffeur d'un pilier. Toutes ces recherches font fimples, d'un ufage journalier ; il feroit facile de les étendre, mais je n'ai voulu ici qu'en établir les principes.

I X.

De la preffion des terres, & des revêtemens.

Si l'on fuppofe qu'un triangle CBa rectangle, folide & **Fig. 7.** pefant, eft foutenu fur la ligne Ba par une force A appliquée en F, perpendiculairement à la verticale CB ; qu'en mêmetemps il eft follicité par fa pefanteur φ, & retenu fur la ligne Ba, par fa cohéfion avec cette ligne, & par le frottement. Soit fait $CB \ldots a$, $Ca \ldots x$; $\delta (aa + xx)^{\frac{1}{2}}$ exprimera l'adhérence de la ligne aB ; φ, pefanteur du triangle CBa, égalera $\frac{gax}{2}$, où g exprime la denfité du triangle.

Si l'on décompofe la force A & la force φ fuivant deux directions, l'une parallèle à la ligne Ba, l'autre qui lui foit perpendiculaire, les triangles $\varphi G \delta$. $F \pi p$, qui expriment ces forces décompofées, feront femblables au triangle CaB; l'on aura donc pour ces forces les expreffions fuivantes ,

φG force perpendiculaire à aB dépendante de $\varphi \ldots \ldots \varphi x : (aa + xx)^{\frac{1}{2}}$

$G \delta$ force parallèle à aB dépendante de $\varphi \ldots \ldots \ldots \varphi a : (aa + xx)^{\frac{1}{2}}$

$F \pi$ force perpendiculaire à aB dépendante de $A \ldots \ldots Aa : (aa + xx)^{\frac{1}{2}}$

πp force parallèle à aB dépendante de $A \ldots \ldots \ldots Ax : (aa + xx)^{\frac{1}{2}}$

Si $\frac{1}{n}$ exprime le rapport conſtant du frottement à la preſſion, l'on aura l'effort que fait le triangle pour couler ſur aB, exprimé par $\left[\varphi a - Ax - \frac{\varphi x - Aa}{n} - \delta\,(aa + xx)\right]:$ $(aa + xx)^{\frac{1}{2}}$; dans le cas d'équilibre, cette expreſſion ſera égale à zéro; d'où l'on tire

$$A = \left[\varphi\left(a - \frac{x}{n}\right) - \delta\,(aa + xx)\right] : \left(x + \frac{a}{n}\right).$$

Mais ſi l'on ſuppoſe que la force appliquée en F, vienne à augmenter, au point qu'elle ſoit prête à mettre le même triangle en mouvement ſuivant la direction Ba; pour lors, en nommant A' cette force, l'on aura

$$\left[A'x - \varphi a - \frac{\varphi x - Aa}{n} - \delta(aa + xx)\right] : (aa + xx)^{\frac{1}{2}},$$

pour l'effort ſuivant Ba; d'où l'on tire, dans le cas d'équilibre,

$$A' = \left[\varphi\left(a + \frac{x}{n}\right) + \delta(aa + xx)\right] : \left(x - \frac{a}{n}\right),$$

quantité qui ſeroit infinie ſi x égaloit $\frac{a}{n}$.

L'on peut remarquer, d'après les deux expreſſions précédentes, que la force A ſera toujours plus petite que la quantité $\frac{ga^2}{2}$, & que la force A' ſera toujours plus grande que cette quantité qui exprime la preſſion, lorſque l'adhérence & le frottement deviennent nuls, ou lorſque le triangle eſt ſuppoſé fluide.

Il eſt donc démontré que lorſque la cohéſion & le frottement contribuent à l'état de repos du triangle, que les limites de la force que l'on peut appliquer en F, perpendiculairement à CB, ſans mettre le triangle en mouvement, ſeront compriſes entre A & A'.

X.

Mais ſi l'on remarque, comme on l'a déjà fait dans l'introduction, que dans une maſſe de terres homogènes

l'adhérence est égale dans tous les points, il faut, pour soutenir cette masse indéfinie, que non-seulement la force A
puisse supporter un triangle donné CBa, mais même parmi
toutes les surfaces $CBeg$, terminées par une ligne courbe
quelconque Beg, celle qui, soutenue par son adhérence &
son frottement, & sollicitée par sa pesanteur, produiroit la
plus grande pression; car, d'après cette supposition, il seroit
évident que si l'on appliquoit en F une force qui ne différât
de celle qui seroit suffisante pour soutenir la surface de la
plus grande pression, que d'une quantité très-petite, la masse
des terres ne pourroit se diviser que suivant cette ligne,
toutes les autres parties restant unies par la cohésion & le
frottement. Il faut donc, pour avoir une force A suffisante
pour soutenir toute la masse, chercher parmi toutes les surfaces $CBeg$, celle dont la pression sur la ligne CB est un
maximum. De même, si l'on vouloit déterminer la plus grande
force qui puisse agir en F, sans troubler l'état de repos, il
faudroit chercher une autre courbe $Be'g'$, telle que la force A'
suffisante pour faire couler la surface $CBe'g$ suivant $Be'g'$
soit un *minimum*, & les limites de la force horizontale, que
l'on peut appliquer en F, sans mettre le fluide en mouvement,
seront comprises entre les limites A & A', où A sera un
maximum, & A' un *minimum*.

Ainsi, il résulte que la différence entre la pression des
fluides dont le frottement & la cohésion sont nuls, & de
ceux où ces quantités ne doivent point être négligées, consiste
en ce que dans les premiers, le côté cB du vase qui les
contient ne peut être soutenu que par une seule force, au
lieu que dans les autres, il y a une infinité de forces conte-
nues entre les limites A & A', qui ne troubleront point
l'état de repos.

Comme il ne s'agit ici que de déterminer la moindre
force horizontale que puisse éprouver le revêtement qui
soutient une masse de terre, sans que l'équilibre soit rompu,
je ne chercherai que la force A.

Je supposerai d'abord que la courbe qui produit la plus

grande preffion eft une ligne droite; ce qui eft conforme à l'expérience, qui donne une furface très-approchante de la triangulaire, pour celle qui fe détache lorfque les revêtemens font ébranlés par le poids des terres.

D'après cette fuppofition & les remarques précédentes, il faut donc, parmi tous les triangles CBa, qui ont pour côté invariable CB, & l'angle C droit, chercher celui qui demande la plus grande preffion A pour l'empêcher de gliffer le long de aB. Ainfi, comme nous avons pour un triangle quelconque, $A = \dfrac{\frac{ax}{2}(a - \frac{x}{n} - \delta(aa + xx)}{(x + \frac{a}{n})}$, l'on aura pour le triangle de la plus grande preffion, par les règles *de maximis & minimis.* $\dfrac{dA}{dx} = \dfrac{(\frac{5a}{2n} + \delta).(aa - \frac{2ax}{n} - xx)}{(x + \frac{a}{n})^2} = 0$, & par conféquent $x = -\dfrac{a}{n} + a\sqrt{(1 + \dfrac{1}{nn})}$.

Subftituant cette valeur de x dans l'expreffion de A, l'on aura $A = ma^2 - \delta la$, m & l étant des coëfficiens conftans, où il n'entre que des puiffances de n; cette force A fera fuffifante pour foutenir une maffe indéfinie $CBlg$.

L'on peut conclure de la formule précédente, que l'adhérence n'influe point fur la valeur de x, ou que les dimenfions du triangle qui produit la plus grande preffion, dépendent abfolument du frottement.

Si le frottement eft nul, quelle que foit l'adhérence, le triangle de la plus grande preffion fera ifofcèle, ou celui dont l'angle fera de 45 degrés.

X I.

Dans la formule précédente, $A = ma^2 - \delta la$, fi l'on fait a variable, l'on aura $dA = da(2ma - \delta l)$ qui exprimera la différence des preffions des furfaces indéterminés CBl, $CB'L$; & puifque la verticale CB ne peut pas

pas porter une moindre force que A, la ligne BB' ne pourra point être fuppofée preffée d'une moindre force que dA; ainfi le *momentum* élémentaire de la force A autour du point E, bafe du revêtement, en nommant b la hauteur totale CE, fera $(b - a)(2ma - \delta l)\, da$, & intégrant, l'on aura pour le *momentum* total autour du point E $\frac{m - b^3}{3} - \frac{\delta l bb}{2}$. Il faudra égaler cette quantité au *momentum* de la pefanteur du revêtement pour en déterminer les dimenfions.

Quant à la forme & aux dimenfions des revêtemens, l'on n'a rien de mieux à confulter dans ce genre que *les Recherches fur la figure des digues*, ouvrage que j'ai déjà cité.

EXEMPLE.

Si l'on fuppofe que le frottement foit égal à la preffion, comme dans les terres qui, abandonnées à elles-mêmes, prennent 45 degrés de talus; fi l'on fuppofe l'adhérence nulle; ce qui a lieu dans les terres nouvellement remuées: pour lors on aura $x = -\frac{a}{n} + a\sqrt{(1 + \frac{1}{nn})} = \frac{4}{10}a$, & $A = \frac{3}{35}a^2$; m fera donc égale à $\frac{3}{35}$, & le *momentum* total autour de G fera $\frac{m\,b^3}{3} = \frac{b^3}{35}$ *; ainfi, fi le mur qui foutient les terres étoit fans talus, que fon épaiffeur fut c, & que fa denfité fut la même que celle des terres, l'on auroit $c = \frac{24b}{100}$, un peu moindre que le quart de la hauteur.

Mais fi le revêtement avoit $\frac{1}{6}$ de talus, en nommant c fon épaiffeur au cordon CD, l'on aura, dans le cas d'équilibre,

---

* Dans cet exemple, comme dans ceux qui fuivent, l'on fuppofe que le revêtement $DCEG$ eft folide & indivifible; que fon frottement, exprimé par une fraction de fa maffe, eft plus grand que la pouffée horizontale A; l'on cherche donc feulement quelles doivent être fes dimenfions, pour qu'il ne puiffe point tourner autour de fon point G.

la formule $\dfrac{b^3}{35} = cb\left(\dfrac{c}{2} + \dfrac{b}{6}\right) + \dfrac{2b^3}{12.3.6}$; d'où

l'on tire à peu-près $c = \dfrac{b}{10}$. Si l'on vouloit augmenter la maffe de la maçonnerie d'un quart en fus de celle qui feroit néceffaire pour l'équilibre, l'on trouveroit $c = \dfrac{b}{7}$; en forte que fi l'on avoit 35 pieds de hauteur de terres à foutenir, il faudroit faire $CD = 5$ pieds; ce qui donne les dimenfions ufitées dans ce cas par la pratique. Je crois la quantité $c = \dfrac{b}{7}$ fuffifante dans l'exécution; d'autant plus, qu'outre l'augmentation de folidité, d'un quart en fus de celle qu'exige l'équilibre, l'on a négligé le frottement qu'éprouve le revêtement, lorfque dans l'inftant de rupture les terres font prêtes à couler le long de GE, ce qui diminue en même-temps la force A & augmente le *momentum* du revêtement.

M. le maréchal de Vauban, dans prefque toutes les places qu'il a fait conftruire, a donné 5 pieds de largeur au cordon, fur $\frac{1}{5}$ de talud. Comme les revêtemens conftruits par cet homme célèbre, paffent rarement 40 pieds, fa pratique fe trouve dans ce cas affez d'accord avec notre dernière formule. Il eft vrai cependant que M. de Vauban ajoute des contre-forts à fes murs; mais cette augmentation de folidité ne doit point être regardée comme fuperflue dans les fortifications, dont les enveloppes ne doivent point être culbutées par le premier coup de canon.

Il réfulte de cette théorie, que dans les terres homogènes, nouvellement remuées, les épaiffeurs des murs qui les foutiennent, mefurées au cordon CD, font comme les hauteurs CE; ce qui paroît devoir diminuer l'épaiffeur que l'on donne ordinairement aux revêtemens qui n'ont que quinze à vingt pieds de hauteur.

X I I.

Dans les terres dont la cohéfion eft donnée, l'on tire de

la formule $A = m a^2 - \delta l a$, qui exprime la preffion des terres, un réfultat affez utile dans leur excavation. Je fuppofe qu'il s'agit de déterminer jufqu'à quelle profondeur l'on peut creufer un foffé, en coupant les terres fuivant un plan vertical, fans qu'elles s'éboulent; car, puifque l'on a, en général, $A = m a a - \delta l a$, fi l'on fait $A = 0$, l'on aura $a = \dfrac{\delta l}{m}$, qui exprimera cette hauteur.

Par des principes analogues, fi la hauteur de l'excavation étoit donnée, l'on trouveroit l'angle fous lequel il faudroit couper les terres pour qu'elles fe foutinffent par leur propre cohéfion.

X I I I.

Si la maffe de terre $c a B$ étoit chargée d'un poids P, il faudroit, dans les formules précédentes, à la place de φ *(art. 10)* fubftituer $P + \dfrac{a x}{2}$, & l'on aura

$$A = \frac{(\frac{g a x}{2} + P).(a a - \frac{x}{n}) - \delta (a a + x x)}{\frac{a}{n} + x},$$

d'où il réfulte

$$x + \frac{a}{n} = \sqrt{[a a (1 + \frac{1}{n n}) - P a (\frac{1}{n n} + 1) : (\frac{g a}{2 n} + \delta)]}.$$

Pour avoir les dimenfions des revêtemens, il faudra fubftituer d'abord cette valeur de x dans la formule qui exprime A, & faire le refte comme dans l'*article 11*.

X I V.

Jufqu'ici nous n'avons point eu égard au frottement qu'é-prouve le triangle $C B a$, en coulant contre $C B$ dans l'inftant de rupture; mais pour peu que l'on y faffe attention l'on voit que ce triangle eft non-feulement retenu par fon frotte-ment fur $B a$, mais encore par le frottement qu'il éprouve en gliffant le long de $c B$, de la part du revêtement; ce dernier frottement pourra être exprimé par $\dfrac{A}{v}$, où $\dfrac{1}{v}$

Z z ij

marque le rapport du frottement & de la preſſion, lorſque les terres font effort pour couler ſur la maçonnerie. Or, dans le cas d'équilibre, le frottement ſur CB, équivaut à une force dirigée ſuivant BC; il faut donc, dans la formule qui donne la valeur de A *(art. 10)*, ſubſtituer à la place de φ

la quantité $(\frac{ax}{2} - \frac{A}{v})$, ce qui donne

$$A = \frac{(\frac{ax}{2} - \frac{A}{v})(a - \frac{x}{n}) - \delta(aa + xx)}{x + \frac{a}{2}} = \frac{\frac{ax}{2}(a - \frac{x}{n}) - \delta(aa + xx)}{a(\frac{1}{n} + \frac{1}{v}) + x(1 - \frac{1}{nv})};$$

d'où l'on tirera, en ſuppoſant que A eſt un *maximum*, & en faiſant $\frac{1}{n} + \frac{1}{v} = m$, & $1 - \frac{1}{nv} = \mu$,

$$x = -\frac{m}{\mu} + \sqrt{\left(\frac{\frac{n}{\mu}(\frac{mga^3}{2} + \delta\mu a^2)}{\frac{ga}{2} + n\delta} + \frac{mma^2}{\mu\mu}\right)}.$$

Subſtituant cette valeur de x dans l'expreſſion de A, & opérant comme ci-deſſus, l'on déterminera les dimenſions des revêtemens.

EXEMPLE.

Si l'adhérence δ eſt ſuppoſée nulle, comme dans les terres nouvellement remuées; ſi le frottement eſt égal à la preſſion, comme dans toutes celles qui prennent 45^d de talus naturel, abandonnées à elles-mêmes; ſi n eſt ſuppoſé égal à v, l'on trouvera pour lors $A = \frac{gx}{4}(a - x)$, & $x = \frac{a}{2}$, l'angle $CBa = 36^d 34'$, & $A = \frac{ga^2}{16}$, le *momentum* total de la preſſion des terres autour du point G ſera $\frac{b^3}{3.16}$; d'où l'on tireroit, pour un mur de terraſſe ſans talus, dont l'épaiſſeur ſeroit c, en ayant égard à la réaction du frottement qui contribue à augmenter le *momentum* de la réſiſtance du revêtement, de la quantité $\frac{c b^2}{16}$, l'équation

$$\frac{b^3}{3.16} = \frac{ccb}{2} + \frac{cb^2}{16};$$

& par conséquent $C = \frac{15}{100} b$, à peu-près; c'est-à-dire qu'un mur de trois pieds de largeur seroit, dans cette hypothèse, en équilibre avec la poussée d'une terrasse de vingt pieds de hauteur.

L'on appliqueroit avec la même facilité les hypothèses de cet exemple à un revêtement qui auroit $\frac{1}{6}$ de talus, comme on le pratique ordinairement dans les murs de terrasse; mais les épaisseurs que donneroit cette application seroient beaucoup plus petites que celle que la pratique semble avoir fixé. Plusieurs causes en effet concourent à faire augmenter les dimensions des revêtemens; en voici quelques-unes.

1.° Le frottement des terres contre la maçonnerie n'est pas aussi fort que celui des terres sur elles-mêmes.

2.° Souvent les eaux filtrant à travers les terres, se rassemblent entre les terres & la maçonnerie & forment des napes d'eau qui substituent la pression d'un fluide sans frottement à la pression des terres; quoique, pour obvier à cet inconvénient, l'on pratique derrière les revêtemens des tuyaux verticaux & des égouts au pied de ces mêmes revêtemens, pour laisser écouler les eaux, ces égouts s'engorgent, ou par les terres que les eaux entraînent, ou par la gelée, & deviennent quelquefois inutiles.

3.° L'humidité change encore non-seulement le poids des terres, mais encore leur frottement. Je puis assurer avoir vu des terres savonneuses, qui se soutenant d'elles-mêmes, lorsqu'elles étoient sèches, sur une inclinaison de 45 degrés, avoient de la peine, quand elles étoient mouillées, à se soutenir sur une inclinaison de 60 à 70 degrés. Enfin, il faut pour que l'on puisse compter sur les dimensions fixées par les formules, que l'eau ne pénètre point les terres dont on cherche la pression, ou qu'en les pénétrant, elle en augmente peu le volume. Cette augmentation de volume que l'humidité procure aux terres, & dont nous avons un exemple dans les lézardes que la sécheresse occasionne à la surface de nos campagnes, produit contre les revêtemens une pression que l'expérience seule peut déterminer.

Ces remarques sont encore indépendantes de la bonté de la maçonnerie, qu'il faut toujours laisser sécher avec soin avant de la charger : elles sont encore indépendantes de la gelée, ennemi sans contredit le plus dangereux dont les maçonneries aient à craindre les effets : car, outre l'augmentation de pression que la gelée produit dans les terres humides, par l'augmentation de volume, outre les engorgemens des tuyaux d'écoulement, l'on peut être sûr que tout mur qui éprouvera de fortes gelées avant d'être sec, perdra nécessairement la plus grande partie de son adhérence, & sera incapable de résistance.

Malgré toutes ces remarques, qui paroissent conduire à conclure qu'il faut des dimensions particulières aux revêtemens, suivant la nature des remblais dont ils éprouvent la pression ; que dans les pays secs & chauds il y a moins d'inconvénient à diminuer les murs de terrasse, que dans les pays humides & froids ; je crois cependant que dans toutes les espèces de terres l'on pourra sans danger fixer les revêtemens à $\frac{1}{6}$ de talus, sur le septième de la hauteur, pour l'épaisseur au cordon (conformément à l'*article 11*).

X V.

De la surface de plus grande pression dans les fluides cohérens.

Jusqu'ici nous avons supposé que la surface qui produit la plus grande pression étoit une surface triangulaire ; la simplicité des résultats que donne cette supposition, la facilité de leur application à la pratique, le desir d'être utile & entendu des Artistes, sont les raisons qui nous ont décidé ; mais si l'on vouloit déterminer d'une manière exacte la surface courbe qui produit la plus grande preſſion : voici, je crois comme on pourroit s'y prendre.

Que CBg, repréſente la ſurface courbe qui produit la plus ₋ Fig. 8. grande preſſion ſur CB, le frottement des terres & la cohéſion étant ſuppoſés les mêmes que ceux du fluide indéfini $gCBl$. Si l'on prend un portion de la ſurface CBg, comme PMg, il eſt évident que cette portion PMg ſera de toutes les ſurfaces que l'on peut conſtruire ſur PM, celle qui produiroit ſur cette ligne la plus grande preſſion ; mais pour avoir la valeur de cette preſſion l'on verra que dans le moment où l'équilibre eſt prêt à ſe rompre, cette ſurface PGM eſt ſoutenue par ſon frottement & ſa cohéſion ſur gM, ſon frottement & ſa cohéſion ſur PM, & ſollicitée par ſa peſanteur φ. Ce que l'on dit par rapport à la portion PMg, on peut le dire par rapport à la portion $P'M'g$. Or, comme dans l'inſtant de rupture, toute la maſſe eſt en équilibre il s'enſuit qu'une portion $PMP'M'$, ſoit élémentaire ou non, ſollicitée par ſa peſanteur, & retenue par ſes frottemens, ſa cohéſion, & les différentes preſſions qu'elle éprouve de la part du fluide qui l'entoure, doit auſſi être en équilibre ; mais pour peu que l'on y faſſe attention l'on remarquera qu'une maſſe PMg ne peut être retenue par ſon adhérence & ſon frottement qui l'empêche de gliſſer le long de PM, ſans que le même frottement & la même adhérence n'agiſſe par ſa réaction ſur la maſſe $CBPM$, dans le ſens contraire. Ainſi en nommant A la preſſion horizontale qu'éprouve la ligne PM, & A' celle qu'éprouve la ligne PM' ; un élément quelconque $PMP'M'$, qui doit être en équilibre, ſera retenu ſuivant une ligne horizontale Fe par la preſſion $(A' - A)$, ſera ſollicité ſuivant la ligne verticale PM, par la réaction du frottement exprimé par $\frac{A}{n}$, par la réaction de l'adhérence δPM, & retenu par le frottement & la cohéſion de $P'M'$, par le frottement & la cohéſion de MM' ; l'on peut donc regarder cette ſurface élémentaire $PP'MM'$, comme un triangle $MM'q$, chargé du poids de l'élément ſollicité par toutes les forces verticales que nous venons de détailler. Soit fait

$$gP\ldots\ldots\ldots x \qquad\qquad PM\ldots\ldots\ldots y,$$
$$gP'\ldots\ldots\ldots x' \qquad\qquad P'M'\ldots\ldots\ldots y',$$
$$gP''\ldots\ldots\ldots x'' \qquad\qquad P''M''\ldots\ldots\ldots y''.$$

Nous avons trouvé *(art. 9 & 10)* qu'une furface triangulaire dont a feroit le côté vertical & x le côté horizontal, follicitée par une puiffance verticale φ, donneroit la preffion

horizontale $A = \dfrac{\varphi(a - \frac{x}{n})}{x + \frac{a}{n}} - \dfrac{\delta(aa + xx)}{x + \frac{a}{n}}$; en com-

parant cette équation avec celle qui auroit lieu pour l'élément $PP'MM'$, l'on trouvera que A repréfente $(A' - A)$; que $\varphi = y.(x' - x) + \dfrac{A - A'}{n} + \delta(y - y')$; que $a = (y' - y)$ & $x = (x' - x)$; ainfi l'équation qui exprime l'état d'équilibre deviendra

$$(A' - A) = \left[y(x'-x) + \frac{A-A'}{n} + \delta(y-y')\right]\left(\frac{y'-y-(\frac{x'-x}{x})}{x'-x+\frac{y'-y}{n}}\right).$$

Suppofons, pour fimplifier, $\delta = 0$, ce qui a lieu pour les terres nouvellement remuées, l'on aura

$$A' - A = \frac{y(x'-x)(y'-y) - (\frac{x'-x}{n})}{\frac{2(y'-y)}{n} + (x'-x)(1 - \frac{1}{nn})}.$$

Par le même raifonnement, l'on trouvera

$$A'' - A' = \frac{y'(x''-x')(y''-y' - \frac{(x''-x')}{n})}{\frac{2(y'-y')}{n} + (x''-x')(1 - \frac{1}{nn})};$$

& par conféquent, en ajoutant enfemble ces deux équations, l'on aura

$$A'' - A = \frac{y(x'-x)(y'-y-\frac{(x'-x)}{n})}{\frac{2(y'-y)}{2} + (x'-x)(1-\frac{1}{nn})} + \frac{y'(x''-x')(y''-y'-\frac{(x''-x')}{n})}{\frac{2(y'-y')}{n} + (x''-x')(1-\frac{1}{nn})};$$

mais

mais puifque la preffion horizontale de la furface PMg doit être un *maximum*, de même que la preffion horizontale de la furface $P''M''g$, il fuit que les quantités y, y', y'' & x, x'' reftant conftantes, x', feul variable, doit être tel qu'il rende $A'' - A'$ un *maximum*; ce qui donne, en différentiant & faifant $y' - y = y'' - y' = dy$, $\dfrac{d(A''-A)}{dx'}$

$$= \frac{y(dy - \frac{(x'-x)}{n})}{\frac{2\,dy}{n} + (x'-x)(1 - \frac{1}{nn})} - \frac{\frac{y}{n}(x'-x)}{\frac{2\,dy}{n} + (x'-x)(1 - \frac{nn}{1})} - \frac{(1 - \frac{1}{nn})y(x'-x)(dy - \frac{(x'-x)}{n})}{[\frac{2\,dy}{n} + (x-x')(x - \frac{1}{nn})]^2}$$

$$- \frac{y'(dy - \frac{(x''-x)}{n})}{\frac{2\,dy}{n} + (x''-x')(1 - \frac{1}{nn})} + \frac{\frac{y'}{n}(x''-x')}{\frac{2\,dy}{n} + (x''-x')(1 - \frac{1}{nn})} + \frac{(1 - \frac{1}{nn})y'(x''-x')(dy - \frac{(x''-x')}{n})}{[\frac{2\,dy}{n} + (x''-x)(1 - \frac{1}{nn})]^2}$$

mais comme les différentes parties correfpondantes de cette équation font des fonctions confécutives femblables, il fuit, en intégrant & fubftituant dx à la place de $x' - x$,

$$B = \frac{y(dy - \frac{dx}{n})}{2(dy + dx(1 - \frac{1}{nn}))} - \frac{y\frac{dx}{n}}{\frac{2\,dy}{n} + dx(1 - \frac{1}{nn})} - \frac{(1 - \frac{1}{nn})y\,dx(dy - \frac{dx}{n})}{[\frac{2\,dy}{n} + dx(1 - \frac{1}{nn})]^2}.$$

Si dans cette équation l'on fait $\zeta\,dy = dx$, l'on trouvera

$$\frac{y(1 - \frac{\zeta}{n})}{\frac{2}{n} + \zeta(1 - \frac{1}{nn})} - \frac{y\frac{\zeta}{n}}{\frac{2}{n} + \zeta(1 - \frac{1}{nn})} - \frac{(1 - \frac{1}{nn})y\zeta(1 - \frac{\zeta}{n})}{[\frac{2}{n} + \zeta(1 - \frac{1}{nn})]^2} = B.$$

Comme dans cette équation réduite, ζ n'eft élevé qu'à la deuxième puiffance, elle aura la forme fuivante,

$$\zeta\zeta + \frac{F' + G'y}{F + Gy}\zeta + \frac{F'' + G''y}{F + Gy} = 0,$$

& par conféquent,

$$\zeta + \frac{F' + G'y}{2(F + Gy)} = \pm \left[\left(\frac{F' + G'y}{2(F + Gy)}\right)^2 - \frac{F'' + G''y}{F + Gy}\right]^{\frac{1}{2}}; F, F', F'',$$

de même que $G\,G'$ & G'' font des coëfficiens conftans.

Si l'on avoit eu égard à l'adhérence, l'on auroit eu précifément une équation de la même forme, & l'on n'y trouveroit de différence que dans les coëfficiens.

L'on peut conclure de cette dernière recherche, que fi un fluide, dont la cohéfion & le frottement feroient donnés, étoit contenu dans un vafe CBg', la preffion contre la paroi CB feroit la même, quelle que fût la figure de Bg; fi l'on pouvoit y infcrire la furface courbe Beg, qui produiroit un *maximum* dans une maffe de fluide indéfinie; mais fi la courbe Beg, qui produit la plus grande preffion, étoit extérieure au vafe; pour lors, il faudroit déterminer, de toutes les furfaces que l'on pouvoit infcrire de ce vafe, celle qui produiroit la plus grande preffion.

Cependant, il faut remarquer que fi l'adhérence & le frottement du vafe & du fluide étoient plus petits que ceux du fluide avec lui-même; pour lors, il fe pourroit que la preffion du fluide contenu dans le vafe fut plus grande que celle du fluide indéfini. Le développement de ces remarques, de même que l'application des formules qui précèdent, demandent un travail exprès, & m'éloigneroit de la fimplicité que je me fuis prefcrite dans ce Mémoire; j'efpère cependant pouvoir une autre fois traiter cette matière dans la théorie des mines, qui, dépendant en partie des principes que je viens d'expliquer, demande encore la folution de quelques Problèmes affez curieux.

X V I.

Des Voûtes.

Fig. 9.

Soit la courbe FAD, décrite fur l'axe FD; foit une feconde courbe fad, décrite extérieurement à la première; foit divifée la courbe FAM en une infinité de parties MM', & de chaque point M, foit tirée la ligne MM', perpendiculaire à la courbe intérieure en M, où formant avec l'élément MM' un angle fuivant une loi donnée; fi l'on fuppofe les deux lignes FAD, fad, telles qu'une portion quelconque $AaMm$, follicitée par la pefanteur, & retenue par la cohéfion & le frottement, foit en équilibre, l'on aura formé le profil d'une voûte. Si l'on fuppofe enfuite que ce profil fe meut, parallèlement à lui-même, & forme une enveloppe folide, comprife entre le tracé du mouvement des deux courbes, l'équilibre, démontré par rapport à ce profil, fera encore vrai, par rapport à cette enveloppe; & la voûte ainfi formée, fera celle que l'on appelle une *voûte en berceau*. C'eft celle dont je me fuis occupé dans les recherches qui fuivent. Les principes que l'on y explique pourront s'appliquer à toutes les autres efpèces de voûtes.

X V I I.

Des Voûtes dont les joints n'ont ni frottement, ni cohéfion.

Soit *aB* le profil d'une voûte, d'une épaiffeur infiniment Fig. 10. petite, dont les joints foient perpendiculaires à la courbe *aB;* l'on demande la figure de cette voûte, follicitée par des puiffances quelconques.

Que toutes les forces qui agiffent fur la portion *aM* foient décompofées fuivant deux directions, l'une verticale, & l'autre horizontale; que la réfultante de toutes les forces verticales foit Qz, que je nomme φ; que la réfultante de toutes les forces horizontales foit $Q\varphi$, que je nomme π; foit de plus $aP \ldots y$, $PM \ldots x$, $Mq \ldots dx$, $qM' \ldots dy$, il eft évident *(art. 1.er)* que la réfultante de toutes les forces qui agiffent fur la portion *aM* doit être perpendiculaire au joint en *M;* & par *l'article 3*, que toutes les forces qui follicitent cette partie de voûte, étant décompofées fuivant deux directions, l'une verticale & l'autre horizontale, perpendiculaires l'une à l'autre; la fomme des forces, fuivant chaque direction doit être nulle; ainfi, fi l'on nomme *P* la preffion du joint en *M*, & que l'on décompofe cette preffion en deux forces, l'une horizontale $\frac{Pdx}{ds}$, & l'autre verticale $\frac{Pdy}{ds}$, l'on aura les deux équations fuivantes $\frac{Pdx}{ds} = \pi$, & $\frac{Pdy}{ds} = \varphi$, & par conféquent, en divifant l'une par l'autre, pour faire difparoître *P*, l'on aura $\frac{dx}{dy} = \frac{\pi}{\varphi}$;

A a a ij

équation qui exprime la figure d'une voûte, follicitée par des puiffances quelconques.

Cette formule fe trouve exactement la même que celle qui a été déterminée par M. Euler *(dans le troifième volume de l'Académie de Péterfbourg)* pour la figure d'une chaîne, follicitée par des puiffances quelconques. Ce qui doit effectivement arriver; car, en renverfant la courbe, & fubftituant la tenfion à la preffion, la théorie précédente s'applique également à l'un ou l'autre cas, & donne précifément la même expreffion. Au refte, la méthode de M. Euler n'a rien de commun avec celle-ci, que le réfultat.

COROLLAIRE I.er

Si la puiffance horizontale étoit conftante & égale à la preffion en a, & fi la réfultante des forces verticales étoit égale à la pefanteur de la portion de la voûte aM; pour lors, l'on auroit $\dfrac{dx}{dy} = \dfrac{A}{\int p\, ds}$; d'où l'on tirera la valeur de p, fi la courbe eft donnée, & de même l'expreffion de la courbe lorfque la loi de pefanteur p eft donnée.

COROLLAIRE II.

Fig. 11. Si l'épaiffeur de la voûte étoit finie, les mêmes fuppofitions exiftantes, que dans le Corollaire précédent; foit R le rayon de la développée au point M; foit z le joint Mm, l'on aura $MM'mm' = \dfrac{ds\,(2R+z)}{2R}$, & par conféquent $\dfrac{dx}{dy}$

$= \dfrac{A}{\int \frac{z\,ds\,(2R+z)}{2R}}$, d'où $\dfrac{A\,ddy}{dx} = \dfrac{z\,ds\,(2R+z)}{2R}$; mais

$R = \dfrac{ds^3}{ddy.dx}$, & $\dfrac{ddy}{dx} = \dfrac{ds^3}{R\,dx^2}$; ainfi, l'on aura $\dfrac{A\,(ds)^2}{R\,dx^2}$

$= \dfrac{z\,(2R+z)}{2R}$; ce qui donne

$$R + z = \left(RR + \frac{2A\,(ds)^2}{dx^2} \right)^{\frac{1}{2}},$$

équation générale pour une voûte quelconque, dans le système de la pesanteur.

EXEMPLE.

Si la courbe intérieure aMB étoit un cercle dont le rayon fut 1, & qu'on cherchât la valeur de z, il est clair que

$$\frac{ds}{dx} = \frac{MM'}{qM'} = \frac{1}{\cos s}; \text{ ainsi } 1 + z = \left(1 + \frac{2A}{(\cos s)^2}\right)^{\frac{1}{2}}.$$

Si l'on suppose qu'au sommet de la courbe le joint $Aa = b$, l'on aura pour lors cos. $s = 1$, & $A = \frac{2b + bb}{2}$,

REMARQUE Iere

Par cette théorie, je n'ai cherché qu'à remplir la première condition d'équilibre, qui exige que toutes les forces qui agissent sur une portion de voûte $GaMm$, aient leur résultante perpendiculaire au joint Mm; mais il est facile de prouver que l'on a satisfait en même-temps à la deuxième condition, qui demande que cette résultante tombe entre les points M & m; car, puisque la force constante A agit perpendiculairement au joint vertical Ga, en un point quelconque S, il s'ensuit que puisque par la condition d'équilibre que l'on vient de remplir, la ligne des résultantes doit couper tous les joints perpendiculairement, elle formera une courbe parallèle à la ligne intérieure aB; ainsi, dans le cas où la force A seroit appliquée en a, la ligne des pressions seroit exactement la même que aMb.

M. Jacques Bernoulli *(Op. vol. II, p. 1119)* en cherchant la figure d'une voûte dont les voussoirs seroient égaux & très-petits, trouve, par les différentes conditions d'équilibre, deux expressions différentes ; mais une fausse estimation dans les angles de cotangente, a donné lieu à l'erreur de M. Bernoulli, & la remarque en a été déjà faite dans les notes par les Éditeurs de ses Ouvrages.

REMARQUE II.

Il fuit encore de la formule générale

$$R + z = (RR + \frac{2A(ds)^2}{dx^2})^{\frac{1}{2}},$$

que toutes les fois que la voûte aB forme au point B un angle droit avec fon axe EB, parallèle à l'horizon, le joint, dans ce point, devient infini ; ou que ce joint eft l'afymptote de la courbe extérieure CD ; car, puifque dans l'équation fondamentale, ds devient infini par rapport à dx, il fuit que $R + z$ devient auffi une quantité infinie. Ce réfultat fe trouve peu conforme avec ce que nous voyons exécuter tous les jours, puifque dans la pratique, les joints horizontaux, au lieu d'être infinis, font fouvent affez petits. Dans la théorie, en outre, la courbe intérieure étant donnée, la longueur du joint eft toujours une quantité donnée ; quantité cependant que les Architectes varient à l'infini dans l'exécution. Mais le frottement & l'adhérence confervent par leur réfiftance l'équilibre, que la force de la gravité tend à détruire. Nous chercherons dans la fuite la manière de faire entrer dans l'expreffion des voûtes ces nouvelles forces coërcitives ; mais l'on peut en attendant inférer de cette remarque, que dans l'exécution, la théorie qui précède, ne peut être, comme nous l'avons déjà dit dans le Difcours préliminaire, que d'une foible utilité.

COROLLAIRE III.

Si la courbe extérieure, de même que la courbe intérieure étoient données, l'on pourroit déterminer, dans le cas d'équilibre, la direction des joints de la manière fuivante.

Fig. 12. Soit fuppofé, comme plus haut, le joint aG vertical, prolongé indéfiniment en l ; foit qM le joint en M, qui, prolongé, rencontre la verticale al en G ; foit φ le centre de gravité de la partie $aGMq$; foit Sp la direction de la force horizontale conftante A qui rencontre en p une verticale paffant par le centre de gravité φ ; la réfultante de toutes les forces fera exprimée par une ligne pn, qui *(art. 1)* doit être perpendiculaire au joint Mq, & paffer entre les points M & q; foit tiré PM, parallèle à l'axe AB, & foit nommé h l'angle PMC. La courbe aMB étant donnée, de même que la courbe GqD, la pefanteur de la maffe $GaMq$ fera exprimée par une fonction de PM & de h; mais les deux triangles femblables prn, PCM, dont les côtés du premier font proportionnels aux forces qui agiffent fur la portion de voûte $GaMq$, donnent l'analogie fuivante : P pefanteur de la portion de la voûte $GaMq : A$

:: cof. h : fin. h, ou $P = \dfrac{A \cos. h}{\text{fin. } h}$. Nous verrons dans la fuite

quels font les points S entre a & G, où l'on peut appliquer la preffion A, quantité déterminée par l'équation précédente, pour fatisfaire à la deuxième condition d'équilibre; c'eft-à-dire, pour que la réfultante pn paffe toujours entre les points M & q.

E X E M P L E.

Si l'on vouloit déterminer la direction des joints d'une plate-bande d'une épaiffeur conftante & donnée; que $aGBb$ Fig. 13. repréfente cette voûte comprife entre deux lignes droites parallèles. La direction du joint vertical aG, de même que la direction du dernier joint Bb, par lequel la voûte s'appuie fur le mur $BLKo$, étant données, l'on cherche la direction de tous les autres joints MM'; foit $aG = a$, $aM = x$, que la direction du joint MM' rencontre la verticale aG en G, l'on aura $GaMM' = P = ax + \dfrac{a^2 \cos. h}{2 \text{ fin. } h}$,

Subftituant cette valeur de P dans l'équation fondamentale $\dfrac{A \cos. h}{\text{fin. } h} = P$, il en réfulte $ax = \left(A - \dfrac{a^2}{2}\right) \dfrac{\cos. h}{\text{fin. } h}$.

Pour avoir la valeur de la conftante A, foit fuppofé que lorfque $x = aB = b$, $\dfrac{\cos. h}{\text{fin. } h}$ égale C. L'on trouvera

$$A = \frac{2ab + a^2 C}{2C} \; ; \; \text{\& par conféquent} \; x = \frac{b \cos. h}{C \sin. h} \; ; \; \text{d'où}$$

l'on conclud que tous les joints d'une plate-bande paffent par le même point C; ce qui donne une conftruction très-facile.

Pour fatisfaire, dans cet exemple, à la deuxième condition de l'*article I.*^{er}, qui exige que la réfultante des forces qui tiennent en équilibre la portion de voûte $GaMM'$, paffe entre les points M & M'; foit φr une ligne verticale paffant par le centre de gravité de la maffe totale $GaBb$. Si fur le joint bB l'on élève au point B une perpendiculaire Bn, qui rencontre la verticale φr en n, & fi, par ce point n on tire une ligne horizontale ns, le point S, ou le joint vertical Ga fera rencontré par cette ligne, fera le point le plus bas fur le joint Ga, où l'on puiffe appliquer la force A, fans que la plate-bande fe rompe. Ainfi, fi la direction du joint Bb étoit telle, que la ligne Bn rencontrât la verticale φr, en un point n, au-deffus de la ligne Gb, il n'y auroit aucun point fur le joint Ga, où l'on pût appliquer la force A, pour conferver l'équilibre, & la plate-bande fe briferoit néceffairement. Il eft très-facile, d'après ces remarques, de déterminer la limite de l'inclinaifon Bb, lorfque l'épaiffeur Ga eft donnée.

Je crois inutile d'avertir que fi la réfultante Bn, pour la maffe totale, paffe par le point B, la réfultante, pour une maffe particulière $GaMM'$, paffera néceffairement entre M & M', puifque la quantité A reftant conftante, les maffes $GaMM'$ diminuent. Ainfi, dès que l'on a fatisfait à la deuxième condition d'équilibre pour le point B, l'on a néceffairement fatisfait à cette même condition pour un point quelconque M.

XVIII.

X V I I I.

De l'équilibre des voûtes, en ayant égard au frottement &
à la cohéfion.

P R O B L È M E.

Dans une voûte, la courbe intérieure aB, *la courbe extérieure* **Fig. 14.**
G b *étant données, les joints* M m, *perpendiculaires aux élémens*
de la courbe intérieure, feront auffi donnés: l'on demande les
limites de la preffion horizontale en f, *qui foutiendra cette voûte,*
en fuppofant qu'elle foit follicitée par fa propre pefanteur, &
retenue par la cohéfion & le frottement des joints.

Soit prife une portion de cette voûte, telle que G a M m,
foit prolongé m M jufqu'en R; foit nommé l'angle R, h;
foit la force de preffion appliquée en f fur le joint vertical aG,
exprimé par A.

Je fuppofe d'abord la portion G a M m folide, en forte
qu'elle ne puiffe fe divifer que fuivant M m. Il faut donc,
pour que cette portion de voûte foit en équilibre, que la
force A foit telle qu'elle l'empêche de glisser fuivant m M;
mais la force dépendante de A, décompofée fuivant M m,
& dirigée fuivant cette même ligne, eft . . . . A fin. h.
La force parallèle à m M, dépendante de φ . . . . φ cof. h.
La force perpendiculaire à m M, dépendante de A . . . . . . A cof. h.
La force perpendiculaire à m M, dépendante de φ . . . . . . φ fin. h.

Ainfi, l'on aura, en ayant égard au frottement & à

l'adhérence, φ cof. $h - A$ fin. $h \dfrac{-\varphi \,\text{fin.} h - A \,\text{cof.} h}{n} - \delta . Mm,$

pour exprimer l'effort que fait cette portion de voûte pour
glisser felon m M; & dans le cas que A fera feulement
fuffifant pour la foutenir, l'on aura

$$A = \frac{\varphi\,(\text{cof.}\ h - \dfrac{\text{fin.}\ h}{n}) - \delta\,Mm}{\text{fin.}\ h + \dfrac{\text{cof.}\ h}{n}}.$$

Or, comme par fa conftruction, la voûte peut non-feulement

B b b

gliſſer ſur le joint mM, mais même ſur tout autre, il ſuit
que pour que la voûte ne ſe rompe point, A ne doit jamais
être moindre que la quantité $\dfrac{\varphi\,(\text{coſ. } h - \frac{\text{ſin. } h}{n}) - \delta\, Mm}{\text{ſin. } h + \frac{\text{coſ. } h}{n}}$,
quelle que ſoit la valeur de h. Ainſi, ſi l'on prend la valeur
de h, telle qu'elle donne pour A un *maximum*, pour lors la
conſtante A, ainſi déterminée, ſera ſuffiſante pour ſoutenir
toute la voûte.

Je ſuppoſe que $A_{,}$ exprime ce *maximum*.

Si l'on cherchoit à déterminer la force en f, de manière
qu'elle fût prête à faire couler la portion de voûte qui
oppoſeroit la moindre réſiſtance, ſuivant Mm, pour lors,
l'on auroit, dans le cas d'équilibre, pour une portion
quelconque $A = \dfrac{\varphi\,(\text{coſ. } h + \frac{\text{ſin. } h}{n}) + \delta\, Mm}{\text{ſin. } h - \frac{\text{coſ. } h}{n}}$; mais comme
aucune portion de voûte ne doit gliſſer ſur un joint quel-
conque Mm, il faut que A ſoit toujours plus petit que cette
dernière quantité. Ainſi il faut chercher le *minimum* de A
qui exprimera la plus grande force que l'on puiſſe appliquer
en f, ſans rompre la voûte, ſuivant un joint Mm; je ſuppoſe
que A' ſoit ce *minimum*.

Ainſi, comme dans le cas de repos, qui eſt celui que
nous cherchons à fixer, la voûte, en tout ou en partie, ne
doit point gliſſer ſur ſes joints dans aucun ſens, il ſuit que
les limites des forces que l'on peut appliquer en f, ſont
compriſes entre $A_{,}$ & A', ou $A_{,}$ exprime la moindre force
qui puiſſe preſſer le point f, & A' la plus grande force qui
puiſſe preſſer ce même point; d'où l'on peut conclure que
ſi $A_{,}$ eſt plus grand que A', il ne peut y avoir d'équilibre,
puiſque la preſſion en f ne pouvant point être plus grande
que A', ne peut point être non plus plus petite que $A_{,}$, que
nous ſuppoſons plus grand que A'.

Pour ſatisfaire à préſent à la deuxième condition d'équilibre,

il faut que la réfultante gv, de toutes les forces qui agiffent fur la portion de voûte $GaMm$, paffe au-deffus du point M, & au-deffous du point m. Il faut, par conféquent, en nommant B la force qui agit en f, que BMQ foit toujours égal ou plus grand que $\varphi g M - \delta' \iota\iota$ (δ' étant une fraction conftante de la cohéfion du mortier, *art. 7)*; & dans le cas où la réfultante pafferoit par le point M, l'on auroit $B = \dfrac{\varphi g M - \delta' \iota\iota}{MQ}$. Si la quantité B étoit fuppofée plus petite que $\dfrac{\varphi g M - \delta' \iota\iota}{MQ}$, pour lors la réfultante gv pafferoit au-deffous du point M, & la voûte fe romproit. Ainfi, pour avoir la force B, fuffifante pour foutenir toute la voûte, il faut chercher le *maximum* de B d'après l'équation précédente, & ce *maximum* exprimera la plus petite force que l'on puiffe appliquer en f; que A, exprime ce *maximum*.

Comme il faut encore, pour fatisfaire à la deuxième condition, que la réfultante Lv paffe au-deffous du point m, il fuit que Bmq doit être plus petit, ou tout au plus égal à $\varphi L q + \delta' \iota\iota$. Ainfi, d'après l'équation $B = \dfrac{\varphi \cdot g q + \delta' \iota\iota}{mq}$, il faut déterminer la conftante B, telle qu'elle repréfente le *minimum* de $\dfrac{\varphi g q + \delta' \iota\iota}{mq}$; & B', déterminé d'après cette confidération, donnera pour Bmq une quantité égale à $\varphi q g + \delta' \iota\iota$, dans un point feulement, & plus petite dans tous les autres points m, & par conféquent B' exprimera la plus grande force que l'on puiffe fuppofer agir en f; d'où l'on conclud que pour remplir la deuxième condition, la force appliquée en f ne peut point être plus petite que B, ni plus grande que B'. Par conféquent, pour joindre les deux conditions enfemble, fi A, ou B, étoient plus grands que A' ou B', l'équilibre ne pourroit point avoir lieu, & la voûte, dont les dimenfions feroient données, fe romproit néceffairement.

Pour avoir actuellement les vraies limites, il fuffit de prendre entre A, & B, la quantité la plus grande, & entre

A' & B' la quantité la plus petite, en forte que fi $B_{,}$ étoit plus grand que $A_{,}$, & B' plus petit que A', $B_{,}$ & B' feroient les véritables limites des forces que l'on pourroit appliquer en f fans rompre la voûte.

Remarque I.

Le frottement eft fouvent affez confidérable dans les matériaux que l'on emploie à la conftruction des voûtes, pour que les différens vouffoirs ne puiffent point gliffer l'un contre l'autre ; en ce cas, l'on peut négliger la première condition d'équilibre ; & il n'eft plus néceffaire que la réfultante des forces qui agit fur une portion quelconque de voûte foit perpendiculaire aux joints qui la terminent ; mais feulement qu'elle tombe fur ces joints. Ainfi, en négligeant la cohéfion des joints, ce qui doit fe faire dans les voûtes nouvellement conftruites ; il fuffit de chercher le *maximum* de $\frac{\varphi g M}{MQ}$, pour déterminer la force $B_{,}$, & le *minimum* de $\frac{\varphi q g}{mq}$, pour déterminer B' ; l'on doit en outre fuppofer que la force B agit en G, fommet du joint, pour rendre la force $B_{,}$ auffi petite qu'elle puiffe être. Il faut cependant remarquer que lorfqu'on cherche à fixer l'état d'équilibre par cette feconde condition, en fuppofant les forces paffant par les points G & M, il faut fuppofer que ces points font affez éloignés de l'extrémité des joints, pour que l'adhérence des vouffoirs ne permette pas à ces forces d'en rompre les angles ; ce qui fe détermine par les méthodes que nous avons employées en cherchant la force d'un pilier.

Remarque II.

Dans la pratique, il fera toujours plus fimple de déterminer les limites de la force B par tâtonnement, que par des moyens exacts. Je fuppofe, par exemple, que l'on prenne la portion GaM de la voûte, telle que le joint Mm faffe un angle de 45 degrés avec une ligne horizontale ; l'on calculera la

force B_l dans cette fuppofition; l'on cherchera enfuite cette
même force par rapport à un fecond joint, peu diftant du
premier, en s'approchant de la clef; fi cette deuxième force
eft plus grande que la première, l'on fera fûr que l'angle de
rupture de la voûte eft entre la clef & le premier joint;
ainfi, en remontant, par cette même opération, vers cette
clef, l'on déterminera facilement la vraie force B_l. Ce calcul
ne fauroit jamais être bien long, parce que par la propriété
de maximis & minimis, il y aura, vers un point M, où l'on
trouve la limite cherchée B_l, très-peu de variations fur un
affez grand développement de la courbe; & qu'ainfi, pour
déterminer cette force B_l, il ne fera néceffaire que d'avoir
à peu-près le point de rupture M; l'on déterminera par les
mêmes moyens la plus grande force B' que puiffe foutenir
une voûte fans fe rompre. Par conféquent, fi les dimenfions
de la voûte étoient données, comme nous le fuppofons ici,
de même que la hauteur du pied-droit BE, fur lequel elle
porte, l'on déterminera facilement quelle doit être l'épaiffeur
Bb de ce pied-droit, pour que la réfultante de la force B_l,
qui agit en G, & de la pefanteur totale de la voûte & de
fon pied droit paffe entre E & e, ou paffe par le point e;
ce qui fatisfera à la deuxième condition de folidité.

La deftination de ce Mémoire, peut-être déjà trop long,
ne me permet pas d'étendre cette théorie, ni de l'appliquer
à toutes les efpèces de voûtes: ainfi, je me contenterai
d'avoir effayé de donner des moyens exacts, & tels que
je les crois abfolument néceffaires pour conftater l'état de
folidité.

En comparant les principes qui précèdent avec les différentes méthodes d'approximation ufitées dans la pratique,
l'on s'apercevra facilement que leurs auteurs n'ont point affez
diftingué les deux conditions d'équilibre néceffaires pour
l'état de repos. Dans celle, par exemple, que l'on attribue
à M. de la Hire, rapportée par M. Bélidor, & pratiquée
par prefque tous les Artiftes, l'on divife la voûte en trois
parties, & l'on calcule la preffion de la partie fupérieure,

en se conformant à la première condition d'équilibre, & l'on détermine ensuite les dimensions des pieds-droits, par la deuxième condition d'équilibre. Or, pour peu que l'on y fasse attention, l'on verra que si l'on divise la partie supérieure vers la clef, & que l'on suppose que cette voûte se rompe en quatre parties, au lieu de se rompre en trois, la force de pression des parties supérieures sera souvent, dans les voûtes plates, beaucoup plus grande que celle qui se détermine par la méthode de M. de la Hire, & que les dimensions des pieds-droits, fixés par cette méthode, seront souvent insuffisantes.

Pl. I.

Sav. Etrang 1773 Pag. 392. Pl. XV.

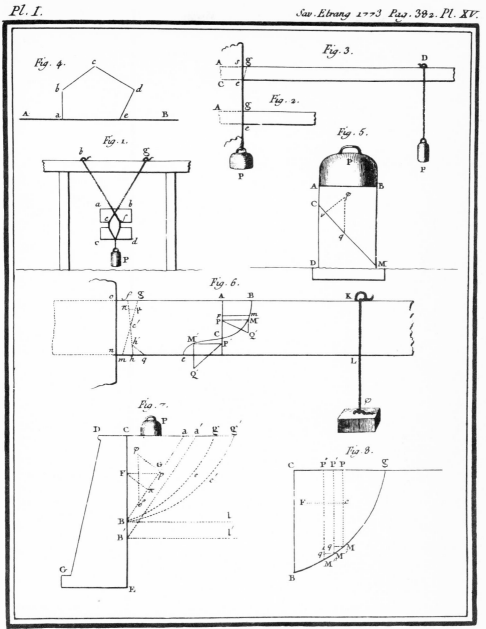

Pl. II.

Sav. Etrang. 1773. Pag. 382. Pl. XVI.

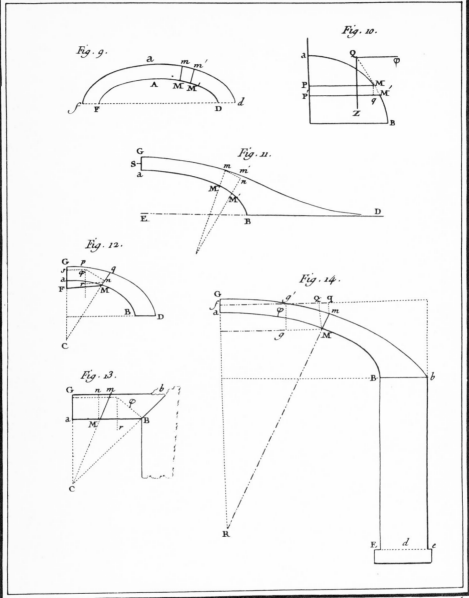

Fig. 9.

Fig. 10.

Fig. 11.

Fig. 12.

Fig. 13.

Fig. 14.

Elis. Haussard Sculp.

I

NOTE[1]

On an application of the rules of maximum and minimum to some statical problems, relevant to architecture

By C. A. COULOMB, Ingénieur du Roi

INTRODUCTION

The object of this paper is to determine, so far as a mixture of calculation and physical principles will allow, the effect of friction and of cohesion in some problems of statics. The following is a brief summary of the subjects treated.

After some preliminary remarks on cohesion, and the reports of some tests on the same subject, the strength of a masonry pier is determined, that is, the weight that it can carry when loaded axially, and the angle at which it must fracture. Since this problem only requires quite simple considerations, which will help in the understanding of all the other sections of this Note, we will try to indicate the principles of its solution.

Suppose that a masonry pier is cut by a plane inclined to the horizontal, in such a way that the two portions of the pier are connected at the cut by a given cohesion, while all the rest of the material is of perfect strength, that is, united by an infinite cohesion. Suppose then that the pier is loaded with a weight; the weight will tend to make the upper portion of the pier slide[2] along the inclined plane dividing it from the lower portion. Thus, in the state of limiting[3] equilibrium, the component of the load resolved parallel to the inclined plane will exactly equal the coherence[4]. If it is now noted that, for a homogeneous pier, the strength is in fact the same everywhere, then, for the pier to carry a given load there must be no section on which the load component can cause the upper portion to slide. Thus, to determine the greatest load that can be carried by a pier, that section must be sought, among all sections, for which the coherence is in equilibrium with a minimum load, since then any load other than that determined from this condition would be too small to fracture the pier[5].

As well as the strength that derives from cohesion, I have considered also that due to friction. The same principles suffice to satisfy both cases, and the results can be applied to all our buildings for which the weight is carried by columns or by some similar means.

Earth pressures against vertical retaining planes are next considered; the method is exactly the same. Consider a right triangular prism, having one face vertical, and the hypotenuse resting on an inclined plane on which the triangle tends to slip; if this triangle, acted upon by its own weight, is supported by a horizontal force, and by its cohesion and its friction acting along the hypotenuse, then the horizontal force necessary to maintain equilibrium can be found easily from the principles of statics. If it is then noted that soils, assumed homogeneous, can separate at rupture not only along a straight line but also along a curve, it follows that to calculate the pressure of an area of soil against a vertical plane, that surface must be sought, among all surfaces in an unbounded vertical plane, which, acted upon by its own weight and resisted by its friction and its cohesion, would require for equilibrium to be resisted by a horizontal force that is a maximum [6]; because it is clear that, all other surfaces requiring a smaller horizontal force for equilibrium, the cohesive mass could not rupture. Since in fact the rupture line is almost straight when soils overturn their retaining walls, it is necessary in practice to seek in an unbounded surface only that triangle, among all triangles loading a vertical plane, which requires the greatest horizontal force for its support. Once this force has been determined then the dimensions of the retaining walls can be found easily.

This section concludes with a method for the exact determination of that surface, of all the curved surfaces which can be constructed within an unbounded medium, for which the pressure against a vertical plane is a maximum, account being taken of cohesion and of friction. This method can be used to find the forces exerted by cohesive media on the walls of their containing vessels.

Lastly this paper investigates the dimensions of arches, their sections of rupture, and the limits which bound their equilibrium state, when cohesion and friction contribute to their strength. Gregory was, I think, the first to show, in *Philosophical Transactions* [7], that, under gravity loading, an arch formed from an infinite number of elements of constant and infinitely small thickness would have the same form as a catenary. I have extended this proposition and I have proved that, whatever the number and the direction of the forces acting upon an arch constructed according to the above assumptions, the form of the arch would be the same as that of a flexible chain acted upon by the same forces. The same principles then suffice to determine the directions of the joints in an arch of finite thickness, or when they must be other than normal to the intrados of the arch. This last case arises for plate-bandes; it is found that, for a given thickness, the joints must be directed to a common centre to ensure equilibrium.

The formulae found by ignoring the friction and cohesion of the joints can be of no use in practice, and this was realized by all the mathematicians who have investigated this subject; so that, to obtain useful results, they were forced to base their calculations on assumptions which might be closer to reality. These assumptions usually consist in taking arches to be divided into several portions, and then in finding the conditions for equilibrium for these different portions. However, since this division is made in a somewhat arbitrary way, I have used the rules of maximum and minimum with the object of determining which would be the true sections of rupture in arches which were too weak, and the limits of the forces that could be applied to those of given dimensions. I have tried as far as possible to make the principles I have used sufficiently clear that a workman with a little learning could understand and use them.

This paper, written several years ago, was originally meant only for my own use, in the different tasks in which I was engaged in my profession; if I dare to present it to this Academy, it is because the weakest work is always received kindly by it if the subject is of practical use. Moreover, the Sciences are memorials dedicated to the public good; each citizen should contribute to them according to his capabilities. While great men, installed in the roof of the building, design and build the upper storeys, ordinary workmen, scattered in the lower storeys, or hidden in the darkness of the foundations, should try only to perfect that which more capable hands have created.

PRELIMINARY PROPOSITIONS

I

Let the plane *abcde*, acted upon by any forces in the same plane, be in equilibrium on the line *AB*; then the resultant of all these forces will be perpendicular to the line *AB*, and will act between the points *a* and *e* [8].   Fig. 4

II

If all the forces acting in the plane are resolved into two directions, one parallel to and the other perpendicular to *AB*, the sum of the components parallel to *AB* will be zero, and the sum of the components perpendicular to *AB* will equal the thrust exerted on the line *AB*.

III

If the thrust exerted on the line *AB* is denoted by *P*, the same plane can be supposed to be acted upon by all the applied forces and also by the reaction to the thrust. But if all these forces, including the reaction to the thrust, are resolved in any two directions at right angles, it follows from equilibrium and from the fact that the two directions are perpendicular that the sum of the components in each direction is zero.

IV

On friction

Friction and cohesion are not active forces like gravity, which always exerts its full effect, but only passive forces; these two forces can be measured by the limits of their strength. For example, when it is stated that, for certain polished woods, friction on a horizontal plane is three pounds for a body weighing nine pounds, this implies that a force less than three pounds will not disturb its equilibrium state.

I will assume here that strength due to friction is proportional to compressive force, as was found by Amontons, although for large bodies friction does not follow exactly this law. According to this assumption, it is found that friction for bricks is three quarters [9] of the compressive force. It will be well to make tests on the materials to be used. It is not possible to state here the friction of stone, since tests made for one quarry will not apply to another.

V

On cohesion

Cohesion is measured by the resistance that solid bodies offer to the simple separation of their parts. Since each element of solids, if they are homogeneous, has this same strength, the total cohesion is proportional to the number of parts to be broken, and hence to the area of rupture. I made some tests to try to find the value of this cohesion, and they gave me the following results.

I took a slab *abcd* of a white stone with a fine and uniform grain;* Test 1
this slab was a foot square and one inch thick. I had it notched Fig. 1
at *e* and *f*, so that *ef* formed a neck two inches wide connecting the
two halves of the slab. I hung the slab by this neck by inserting
two ropes knotted in a sling, and with two other ropes I hung a
balance pan that I loaded with a weight *P*. The weight had to be
increased to 430 lb to break the slab at *ef*, which gives a cohesive force
of 215 lb/in².

I wished to see whether, to break a stone by a force directed along Test 2
the plane of fracture, the same weight was necessary for fracture as in
the previous test, where the force was perpendicular to that plane.
For this I placed the small body *ABCD* in a mortise *AGeg*, and I hung Fig. 2
a pan by the rope *eP* which passed round the body at the section
next to the mortise. The small body was two inches wide and one
inch deep, which gives the same rupture area as in the previous test;
it did not break until the pan was loaded to 440 lb. I repeated this
and the first test several times, and I nearly always found that
a larger force was needed to break the body when the force was
directed along the fracture plane than when it was perpendicular to
that plane. However, since the difference is here only $\frac{1}{44}$ of the total
weight, and was often found to be smaller, I have neglected it in the
following theory.

I wished to see how a body fractured when it was broken by Test 3
a force acting on it with a lever arm. I therefore encastred in a mortise *ACeg* a piece of the same stone as in the previous test, having Fig. 3
1 inch depth and 2 inches width, and being 9 inches in length from
g to *D*, where I hung a weight *P*; this weight was 20 lb when the
body broke at *eg*.

VI

I repeated the same tests on Provence bricks, well-fired and with
a very uniform grain; I found that their cohesion, when broken by
a force perpendicular to the plane of fracture as in the first test, was
between 280 and 300 lb/in². I found also that a mortar made from
four parts of sand to three of lime, and two years old, carried 50 lb/in²
perpendicular to the fracture plane. This last test, made in Martinique,
cannot be generalized. The strength of mortar varies sometimes by
a factor of two or even three, according as the climate is wet or dry,
according to the qualities of the sand, of the lime, and of the stone

* This stone is found near Bordeaux, and is used to construct the facings of the
large buildings of that town.

used in the body of the masonry, and according to the age of that masonry; nothing can be fixed about it, and individual tests must be made in each place.

VII

Remarks on the fracture of bodies

Fig. 6   Suppose a rectangular body *onKL* is arranged as a cantilever, and fixed at *on* with the faces of the body horizontal and vertical; suppose further that the body is cut by a vertical plane, whose trace is *AD*, perpendicular to the face *onKL*, and is loaded by a weight ϕ, attached to its end *L*. It is clear, by considering only one vertical face of the body, all the others being equal and parallel, that all the points of the line *AD* will be stressed in preventing the rupture of the body by the weight ϕ; it follows therefore that an upper part *AC* of the line acts in tension directed along *QP*, while the lower part acts in compression along *Q'P'*. If all these forces, whether tensile or compressive, are resolved in two directions, vertical and horizontal, denoted by *QM* and *PM*, and if through all points *M* is constructed the curve *BMCe*, then this curve will give the locus of all the stresses acting perpendicular to the line *AD*. Thus the portion *ADKL* must be supposed to be acted upon by all the horizontal forces *PM*, by all the vertical forces *MQ*, and by the weight ϕ; it follows from article 3 that since there is equilibrium, the sum of the horizontal forces is zero, and that therefore the total tensile force corresponding to *ABC* must equal the total compressive force *CeD*. Further, by the same article, the sum of the vertical forces *QM* must equal the weight ϕ; and further, by the principles of statics, the sum of the moments about *C* of all the tensile and compressive forces must equal the moment of the weight ϕ about the same point, which gives the equation $\int Pp.MP.CP = \phi LD$. We have to satisfy the three conditions above, therefore, whatever the relation between the deformation and the cohesion of the elements of a body [10].

Suppose, for example, that one wishes to find the weight that can be carried by a piece of perfectly elastic wood, that is, one that compresses or extends, when loaded axially, in proportion to the load which compresses or extends it. Let the element *ofnh*, adjacent to the wall, represent a very small section of the wood in its original state. If the piece of wood is loaded with a weight ϕ, the upper end of the line *fh* will move to *g*, and the lower portion to *m*; the line *fh* [11] will become *gm*. But since, by hypothesis, the tensile stresses, and similarly the compressive stresses, are represented by the lengths $\pi\mu$

of the triangle *fge*, it follows that the compressive triangle *emh* must equal the tensile triangle *fge*. Thus if the stress at the point *f*, represented by *fg*, is denoted by δ, and since *fe* will equal $\frac{1}{2}fh$, then the moment of the small tensile triangle will be $\frac{1}{3}\delta(ef)^2$ which, when added to the moment of the small compressive triangle, must give $\frac{1}{6}\delta(fh)^2 = \phi.nL$. Here δfh, at the instant of fracture, expresses the cohesive strength against a force acting perpendicular to the line *fh*, assuming [12], however, that the stresses *MQ* have very little influence on the strength of solids; this is reasonably correct when the lever arm *nL* of the weight ϕ is much larger than the depth *fh*.

But if it is assumed that the member, when about to break, is composed of stiff fibres, that can be neither compressed nor extended; if it is assumed further that the body fractures by rotation about the point *h*, then each point of the depth *fh* will be subject to the same stress. The point *h* will be subject to a compressive force equal to δfh, and the moment of the small cohesive triangle [13] will be $\frac{1}{2}\delta(fh)^2$. Let us apply this last hypothesis to our tests.

I found in the first test that an area of two inches width and one inch depth gave a strength equal to 430 pounds. In the third test the dimensions are the same, and further, *hL* equals 9 inches. Thus, if the last hypothesis were correct, I should have found $P = \frac{1}{2}(430/9)$, or almost 24 pounds, but the test gave 20 pounds for *P*; it cannot therefore be assumed in the fracture of stone that the fibres are perfectly stiff, nor that the point of rotation is precisely at *h*. This result could have been foreseen from the fairly obvious fact that, in taking *h* as the point of rotation, it would be necessary to limit the stress at *h* to a finite value in order not to destroy the cohesion, which is impossible since the cohesion only has a finite value for a finite area. It is necessary therefore, in the state just before fracture, that the force shift to a point *h'*, such that the coherence of *h'q* is strong enough to resist the force $\delta fh'$ acting on the line *hh'* and resolved in the direction *h'q* [14]. We will give below the way of finding the vertex *q* of the triangle *h'hq*.

The Abbé Bossut in an excellent paper on the form of dykes, a work which combines an imaginative mind, a physicist's wisdom, and a mathematician's precision, seems to have been the first to identify and define the difference found between the fracture of timber and of stone.

Fig. 1

47

VIII

Strength of masonry piers

Let a homogeneous masonry pier, which I shall first take to be square, be loaded with a weight P; required the direction of the line CM along which the pier will fracture, and the magnitude of the weight necessary for that fracture.

Fig. 5

I assume here that the cohesion gives an equal strength whether the force acts parallel or perpendicular to the plane of fracture, as observed in the first and second tests [15]. I assume also that the pier is made of a homogeneous material, having cohesion δ. Take any section CM inclined to the horizontal, and perpendicular to the vertical plane face $ABDM$ of the pier. Supposing for the moment that the coherence of the upper portion $ABCM$, like that of the lower portion CDM, is infinite, it is clear that the bulk of the column would tend to slide along CM; and therefore, if the two portions were united by a cohesive force equal to the natural cohesion of the pier, in order for the column to break along CM the magnitude of the weight P resolved in that direction would have to equal, or be greater than, the coherence of CM. Let the angle at M be x, let $DM = a$, and P be the weight whose magnitude, represented by ϕq, is resolved in the directions ϕr and rq perpendicular and parallel to the line of fracture. If, for simplicity, the weight of the column is neglected, then
$$\delta CM = \delta a / \cos x \quad \text{and} \quad rq = P \sin x,$$
and hence, for limiting equilibrium,
$$P = \delta a^2 / \sin x \cos x.$$
But since the column must be able to carry the weight P without breaking, whatever the section CM, then the weight P must always be less than the quantity $\delta a^2 / \sin x \cos x$, whatever the value of x. This condition will be satisfied if P is determined as a minimum from the equation $P = \delta a^2 / \sin x \cos x$, which gives
$$\mathrm{d}P = \frac{\delta a^2 (-\mathrm{d}x \cos^2 x + \mathrm{d}x \sin^2 x)}{(\sin x \cos x)^2},$$
and hence $\sin x = \cos x$. Thus the greatest weight that the column could support without breaking is equal to $2\delta a^2$, double the strength it would have in tension, and the angle of least strength, or of fracture, will be $45°$.

We have assumed in this analysis that the section represented by CM was perpendicular to the vertical side $ABDM$, but the same results would have been found for any section, provided that it made the same angle with the horizontal. It may be noted that, by the theory of projection, oblique sections of a pier are in a ratio with their horizontal projections as the inverse cosine of the angle between these two planes; then, denoting this angle by x, and by A the area of the base, equal here to a^2, the expression $\delta a^2/\cos x$ is found for the coherence of the oblique section, and $P\sin x$ for the force which tends to make the upper portion of the column slide on the inclined plane acting as its base, no matter where the plane of the section lies. Since these expressions are exactly the same as those above they must therefore give the same results; from which may be concluded that, whatever the shape of the horizontal base of a pier, if the area of that base is constant, its strength will be the same.

IX

In the above solution we have not allowed for the effect of friction which opposes the fracture of the pier. If it is wished to take this into account, then, keeping the previous notation, it would be found that $P\cos x$ is the component of the weight acting on CM, and since the frictional force is proportional to the compressive force, it will be equal to $(P\cos x)/n$, n being a constant quantity. The bulk of the pier $ABCM$, loaded by the weight P, is then supported by cohesion and by friction; thus, increasing the weight until it is sufficient to break the pier, we have

$$\frac{a^2\delta}{\cos x}+\frac{P\cos x}{n}=P\sin x,$$

or

$$P = \delta a^2[\cos x(\sin x - n^{-1}\cos x)]^{-1}.$$

In order to find the weight that the pier can carry without breaking, the value of P must, by the above principles, be made a minimum, which gives $dx[\sin x(\sin x - n^{-1}\cos x)] - dx\cos x(\cos x + n^{-1}\sin x) = 0$, and hence

$$\cos^2 x + 2n^{-1}\sin x\cos x = \sin^2 x;$$

from which

$$\cos x = \sin x[(1+n^{-2})^{\frac{1}{2}}-n^{-1}],$$

that is,

$$\tan x = [(1+n^{-2})^{\frac{1}{2}}-n^{-1}]^{-1}.$$

If the pier were of brick, then (article 4) $1/n = \frac{3}{4}$, $\tan x = 2$, $\sin x = 2\cos x$; hence $\cos x = (\frac{1}{5})^{\frac{1}{2}}$,

$$P = \frac{\delta a^2}{\cos x(2\cos x - \frac{3}{4}\cos x)} = 4\delta a^2,$$

and the angle at M will be $63° 26'$. Thus the compressive force required to break a brick column is four times the tensile force required to break the same column.

Musschenbroek (*Essai de Physique*, French translation, vol. 1, p. 354[*16*]) found that a square brick pier, $11\frac{1}{2}$ inches long with sides measuring 5 *lignes*[*17*], was broken by a load of 195 lb. In Musschenbroek's test, since the sides were $\frac{5}{12}$ inch, the cross-sectional area was $\frac{25}{144}$ in². Now, in article 6, we found that a square inch of brick carries a load of 300 lb perpendicular to the plane of fracture. Thus in this test $\delta a^2 = 300(\frac{25}{144}) = 52$ lb, which gives the tensile strength; but since $P = 4\delta a^2$, it follows from our theory and our tests that this physicist should have found 208 lb, which is little different from the value of 195 lb[*18*] of his test.

I must point out that the way in which Musschenbroek finds the strength of a masonry pier has no connection with that which I have used. A pier, loaded by an axial compressive force, only breaks, says this eminent physicist, because it starts to bend; otherwise it would carry any magnitude of load. Starting from this principle, he finds the strength of square piers to be inversely proportional to the square of their lengths, and proportional to the cube of their sides. Thus if the pier for which we have just calculated the strength had had only half its original length, it should have carried a load four times as great, that is, 832 lb; in fact I think I have shown that it would have carried only the same weight of 208 lb.

It may be concluded from the formula that the strengths of homogeneous piers are proportional to their cross-sections.

The fracture angle of an incompressible column, loaded by a force inclined to its horizontal base, could be found by the same principles, provided that the line of action of the force falls within that base; if it should fall outside the base, there would be some other considerations which would make the solution of this problem a little more difficult.

The above principles can also be used to determine the height to which a tower can be built without crushing under its own weight. Suppose, for simplicity, that that height is much larger than the width, so that the small prism CDM can be neglected. Instead of the quantity P, the mass of a tower which has the same weight must be substituted in the formulae. Let us assume, for example, that it is built of brick; since a cubic foot of brick weighs about 144 lb, a small prism of one square inch cross-section, and a foot high, would weigh one pound. Thus, since an area of one square inch can carry a tensile force of 300 lb, and a compressive force of twice that value if friction is neglected, it is clear that if the tower is considered to be made up of a collection of small prisms, each of one square inch cross-section and of 600 feet height, its cohesion would enable it to stand. If friction were allowed for, the tower could be built, by the same principles, to a height of 1200 ft; if the tower were replaced by a pyramid, it could be built three times as high.

If the tower were supported on several piers, the height to which it could be built would be in direct proportion to the cross-sectional area of those piers; so that if, for example, the cross-sectional area of the piers were one-sixth of the cross-sectional area of the tower, it could not rise to a height greater than 100 ft above the columns, if friction is neglected, or 200 ft, allowing for friction. The weight of the piers is neglected here; it would be easy to take it into account.

When several vaults spring from the same pier, and mutually buttress and support each other against horizontal thrust, then, since the resultant of their actions is vertical, and directed along the axis of the pier, it would be easy to find by this method the thickness of the pier. All these calculations are simple, and of practical use; it would be easy to extend them, but I wished here only to establish the principles.

IX (*bis*)
On earth pressures, and retaining walls

Let a heavy triangular right-angled prism CBa be supported on the line Ba by a force A acting through F perpendicular to the vertical line CB; suppose also that it is acted upon by its own weight ϕ, and restrained along the line Ba by its cohesion with that line and by friction. Let $CB = a$, $Ca = x$; then $\delta(a^2 + x^2)^{\frac{1}{2}}$ gives the coherence of the line aB. The weight ϕ of the triangle CBa will equal $gax/2$, where g denotes the density of the triangle. Fig. 7

If the forces A and ϕ are resolved in two directions, parallel and perpendicular to the line Ba, the triangles $\phi G\delta$ and $F\pi p$, which represent these resolved forces, will be similar to the triangle CaB. Thus the forces will be given by the following expressions:

ϕG, force perpendicular to aB, component of ϕ: $\quad \phi x/(a^2 + x^2)^{\frac{1}{2}}$,

$G\delta$, force parallel to aB, component of ϕ: $\quad \phi a/(a^2 + x^2)^{\frac{1}{2}}$,

$F\pi$, force perpendicular to aB, component of A: $\quad Aa/(a^2 + x^2)^{\frac{1}{2}}$,

πp, force parallel to aB, component of A: $\quad Ax/(a^2 + x^2)^{\frac{1}{2}}$.

If $1/n$ denotes the constant coefficient of friction, then the component of force tending to make the triangle slide along aB will be expressed by

$$[\phi a - Ax - (\phi x + Aa)\, n^{-1} - \delta(a^2 + x^2)]/(a^2 + x^2)^{\frac{1}{2}}\ [19];$$

51

for equilibrium, this expression will be equal to zero, from which

$$A = [\phi(a - xn^{-1}) - \delta(a^2 + x^2)]/(x + an^{-1}).$$

But if it be assumed that the force applied at F increases to a value such that the same triangle is on the point of moving in the direction Ba, then, denoting that force by A', the force along Ba will be

$$[A'x - \phi a - (\phi x + A'a)\, n^{-1} - \delta(a^2 + x^2)]/(a^2 + x^2)^{\frac{1}{2}};$$

from which, for equilibrium,

$$A' = [\phi(a + xn^{-1}) + \delta(a^2 + x^2)]/(x - an^{-1}),$$

an expression which becomes infinite when $x = a/n$.

It may be noted from the above two expressions that the force A will always be less than $ga^2/2$, and that the force A' will always be greater than the same quantity, which gives the value of the thrust when the cohesion and the friction become zero, that is when the triangle is assumed to be a fluid [20].

It has thus been shown that when cohesion and friction contribute to the stability of the triangle, then the limits of the force which can be applied at F perpendicular to CB without moving the triangle will lie between A and A'.

X

But if it is noted, as has already been done in the introduction, that in a mass of homogeneous soil the cohesion is the same at all points, then, to sustain this unbounded mass, the force A must be capable of supporting not only a given triangle CBa, but also that surface, among all the surfaces $CBeg$ bounded by any curve Beg, which, supported by its cohesion and its friction and acted upon by its own weight, would exert the largest thrust. It will be clear from this statement that if a force were applied at F infinitesimally smaller than that required to support the surface of greatest thrust, the mass of soil could cleave only along that curve, while all the other portions remained united by cohesion and by friction. Thus, to find the force A sufficient to support the whole mass, that surface must be sought, among all the surfaces $CBeg$, for which the thrust on the line CB is a maximum. Similarly, to find the greatest force which can act through F without disturbing equilibrium, another curve $Be'g'$ must be sought such that the force A' sufficient to cause the surface $CBe'g'$ to slide along $Be'g'$ is a minimum. The limits of the horizontal force that can be applied at F without moving the soil will be bounded by the limits A and A', where A will be a maximum and A' a minimum [21].

Thus it follows that the difference between forces in fluids for which friction and cohesion are zero and those for which these quantities cannot be neglected is that for the former, the side *CB* of the vessel containing them can be supported only by a unique force, while for the latter there is an infinite number of forces lying between the limits of *A* and *A'* which will not disturb equilibrium.

Since the question here is only to find the least horizontal thrust which must be exerted on a retaining wall which supports a mass of soil in equilibrium, I will only determine the force *A*.

I will assume first that the curve which gives the greatest thrust is a straight line; experience shows that when retaining walls are overturned by earth pressures, the surface which breaks away is very close to triangular.

From this assumption and the above remarks, that triangle must be sought, among all the triangles *CBa* which have a constant side *CB* and a right angle at *C*, which requires the largest force *A* to prevent slip along *aB*. Thus, since we have for any triangle,

$$A = \frac{\frac{1}{2}gax(a - xn^{-1}) - \delta(a^2 + x^2)}{(x + an^{-1})} \ [22],$$

the triangle of greatest thrust will be given by, from the rules of maximum and minimum,

$$\frac{dA}{dx} = \frac{(\frac{1}{2}gan^{-1} + \delta)(a^2 - 2axn^{-1} - x^2)}{(x + an^{-1})^2} = 0,$$

and hence $x = -an^{-1} + a(1 + n^{-2})^{\frac{1}{2}}$. On substituting this value of x into the expression for *A*, it will be found that $A = ma^2 - \delta la$, where *m* and *l* [23] are constant coefficients containing only powers of *n*; this force *A* will be large enough to support an indefinite mass *CBlg*.

It may be deduced from the above formula that cohesion does not influence the value of x, and that the dimensions of the triangle which gives the greatest force depend only on friction.

If the friction is zero, then, whatever the cohesion, the triangle of greatest force is isosceles, of base angle 45°.

XI

In the above formula, $A = ma^2 - \delta la$, if a is considered as a variable, then $dA = da(2ma - \delta l)$, which expresses the difference of thrust between two unbounded surfaces *CBl*, *CB'l'*. Since the vertical *CB* cannot carry a force less than *A* [24], the line *BB'* cannot be considered to carry a force less than dA. Thus the moment of the elementary force dA about the point E, the base of the retaining wall whose total height CE is b, will be $(b-a)$ $(2ma-\delta l)$ da, and, integrating, the total moment about E will be $\frac{1}{3}mb^3 - \frac{1}{2}\delta lb^2$. This quantity must be equated to the moment of the weight of the retaining wall to determine its dimensions.

As to the shape and dimensions of retaining walls, there is no better reference on this subject than *les Recherches sur la figure des digues*, the work I have already cited.

EXAMPLE

Assuming that the coefficient of friction is unity, as for soils which take a slope of 45° when left to themselves, and that the cohesion is zero, as for newly-turned soils, then

$$x = -an^{-1} + a(1 + n^{-2})^{\frac{1}{2}} = \tfrac{4}{10}a,$$

and $A = \frac{3}{35}ga^2$ [25]; m will thus be equal to $\frac{3}{35}g$, and the total moment about G will be $\frac{1}{3}mb^3 = \frac{1}{35}gb^3$.* Thus if the wall which retains the soil is without batter and has thickness c, and has the same density as the soil, then $c = \frac{24}{100}b$ [26], a little less than a quarter of the height.

But if the retaining wall has a batter of 1 in 6, then, denoting by c the thickness at the ridge CD, for limiting equilibrium

$$\frac{b^3}{35} = cb\left(\frac{c}{2} + \frac{b}{6}\right) + \frac{2b^3}{12.3.6},$$

from which, approximately, $c = \frac{1}{10}b$. If it is desired to increase the mass of the masonry by a quarter above that which would be necessary for equilibrium, then $c = \frac{1}{7}b$. Thus if a height of 35 ft of soil is to be supported, then CD must be made equal to 5 ft, which gives the dimensions used for this case in practice. I think that $c = \frac{1}{7}b$ is sufficient in practice, more especially since, not only has the mass been increased by a quarter above that required for equilibrium, but also the friction exerted upon the retaining wall has been neglected. At the moment of rupture, the soil is about to slide along CE [28], which diminishes A and at the same time increases the moment of the retaining wall.

Marshal Vauban, in almost all the fortresses he built, made the ridge 5 ft wide, with a batter of $\frac{1}{5}$. Since the retaining walls built by

* In this example, as in the ones following, the retaining wall $DCEG$ is assumed to be of infinite strength, and the friction, calculated as a fraction of the weight, is greater than the horizontal thrust A [27]; thus the only problem is to find its dimensions so that it cannot overturn about its toe G.

this famous man were rarely higher than 40 ft, his practice is in this case in reasonable agreement with our last formula. It is true however that Vauban added buttresses to his walls; but this strengthening should not be thought superfluous in fortifications, of which the enceintes ought not to fall to the first cannon ball.

It follows from this theory that, for newly-turned homogeneous soils, the thicknesses of the walls supporting them, measured at the ridge CD, are proportional to the heights CE; it would therefore seem that the usual thickness given to retaining walls of only fifteen or twenty feet height ought to be reduced.

XII

For soils of known cohesion, the formula $A = ma^2 - \delta la$ giving the soil thrust leads to a useful result for excavations. Suppose that the depth is required to which a trench can be dug, the walls being cut vertically, without them falling in; now, since in general $A = ma^2 - \delta la$, if $A = 0$ then $a = \delta l/m$, which will give that depth.

By similar principles, if the depth of the excavation were given, the angle would be found at which the soil should be cut so that it would support itself by its own cohesion.

XIII

If the mass of soil CaB were loaded with a weight P then ϕ (article 10) should be replaced in the above formulae by $P + \frac{1}{2}gax$, and

$$A = \frac{(\frac{1}{2}gax + P)(a - xn^{-1}) - \delta(a^2 + x^2)}{an^{-1} + x}$$

from which

$$x + an^{-1} = [a^2(1 + n^{-2}) - Pa(1 + n^{-2})/(\tfrac{1}{2}gan^{-1} + \delta)]^{\frac{1}{2}}.$$

To obtain the dimensions of the retaining walls, this value of x must first be substituted in the expression for A, and then the procedure of article 11 must be followed.

XIV

So far we have not considered the friction exerted on the triangle CBa in sliding along CB at the instant of rupture. As soon as this is noted it becomes obvious that the triangle is supported not only by its friction on Ba, but also by the friction that is exerted on it by the retaining wall in sliding along *CB*. This latter friction can be expressed by A/ν, where $1/\nu$ denotes the coefficient of friction when soil attempts to slip over masonry. Now, for limiting equilibrium, the friction force on *CB* is equal to a force directed along *BC*; thus, in the formula giving the value of *A* (article 10), the quantity $(\frac{1}{2}gax - A\nu^{-1})$ must be substituted for ϕ, which gives

$$A = \frac{(\frac{1}{2}gax - A\nu^{-1})(a - xn^{-1}) - \delta(a^2 + x^2)}{x + an^{-1}} = \frac{\frac{1}{2}gax(a - xn^{-1}) - \delta(a^2 + x^2)}{a(n^{-1} + \nu^{-1}) + x(1 - n^{-1}\nu^{-1})}$$

from which, assuming *A* to be a maximum, and setting $n^{-1} + \nu^{-1} = m$ and $1 - n^{-1}\nu^{-1} = \mu$,

$$x = -ma\mu^{-1} + \left[\frac{n\mu^{-1}(\frac{1}{2}mga^3 + \delta\mu a^2)}{\frac{1}{2}ga + n\delta} + m^2a^2\mu^{-2}\right]^{\frac{1}{2}}.$$

On substituting this value of *x* into the expression for *A*, and proceeding as above, the dimensions of the retaining walls will be found.

EXAMPLE

If the cohesion δ is assumed to be zero, as for newly-turned soils, if the coefficient of friction is unity, as in all soils which have a natural slope of $45°$ if left to themselves, and if *n* is taken equal to ν, then

$$A = \tfrac{1}{4}gx(a - x), \quad x = \tfrac{1}{2}a, \quad \angle CBa = 26° \, 34', \quad A = \tfrac{1}{16}ga^2,$$

and the total moment of the soil thrust about the point *G* is $gb^3/3.16$. Thus, for a terrace wall without batter, and of thickness *c*, the friction force contributes to an increase in the moment of resistance of the retaining wall by an amount $\tfrac{1}{16}gcb^2$, so that

$$\frac{b^3}{3.16} = \frac{c^2b}{2} + \frac{cb^2}{16}.$$

Hence, very nearly, $c = \tfrac{15}{100}b$; that is, a wall three feet thick would be stable, according to this hypothesis, under the thrust of a terrace twenty feet high.

The hypotheses of this example could be applied equally easily to a retaining wall with a batter of $\tfrac{1}{6}$, the usual value in practice for terrace walls, but the thicknesses resulting from this calculation would be much smaller than those that practice seems to have fixed. In fact several causes contribute to increase the dimensions of retaining walls; here are some of them.

1. Friction of soil against masonry is not so large as internal friction of soil.

2. Often water, percolating through the soil, collects between the soil and the masonry and forms a layer of water, replacing soil forces by the frictionless pressure of a fluid. Even though, to avoid this problem, vertical pipes are placed in practice behind retaining walls, and drains at the feet of the same walls, so that the water can run off, these drains get blocked, either by soil carried along by the water, or by ice, and sometimes become useless.

3. Moisture content affects not only the weight of soils, but also their friction. I have myself seen Fuller's earth which, when dry, has an angle of repose of 45°, scarcely maintaining an angle of 60° to 70° [29] when wet. Thus, if the dimensions determined from the formulae are to be relied upon, water must not penetrate the soils whose forces are being sought, or, if it does penetrate, it must increase the volume only slightly. This increase of volume that moisture causes in soil, and of which we have an example in the cracks that drought causes in the surface of our fields, causes a thrust on retaining walls that can only be determined by experiment.

These remarks have not yet taken into account the quality of the masonry, which must always be left to dry carefully before being loaded. They have not yet taken into account frost, the enemy whose effects are without doubt the most dangerous that masonry has to fear; since, as well as the increase of pressure that frost produces in wet soils by the increase of volume, and as well as the blockage of the drainage pipes, it is certain that any wall which experiences severe frosts before being dry will necessarily lose the greatest part of its cohesion, and will have no strength.

Despite all these remarks, which seem to lead to the conclusion that retaining walls should be given individual dimensions according to the nature of the backfill which exerts a thrust on them, and that there is less danger in decreasing the size of terrace walls in dry hot countries than in damp cold countries, nevertheless I think that, for all kinds of soil, retaining walls can be designed without danger with a batter of $\frac{1}{6}$, and with the ridge one seventh of the height (as in article 11).

XV

On the surface of greatest thrust in cohesive fluids [30]

Up to now we have assumed that the surface which gives the greatest thrust was a triangular surface. The reasons for adopting this assumption were the simplicity of the results obtained, the ease of their application in practice, and the wish to be useful to and understood by workmen. But if the curved surface which gives the greatest thrust is to be determined exactly, here, I think, is the way to go about it.

Fig. 8   Let *CBg* represent the curved surface which produces the greatest thrust on *CB*, assuming the soil friction and cohesion to be the same as those of the unbounded fluid *gCBl*. If a portion of the surface *CBg* is considered, say *PMg*, it is clear that the portion *PMg* will be the surface, of all the surfaces that can be constructed on *PM*, which would give the greatest thrust on that line. To obtain the value of this thrust it will be seen that in the limiting equilibrium state, the surface *PgM* is supported by its friction and its cohesion along *gM* and its friction and its cohesion along *PM*[31], and is loaded by its own weight ϕ. This statement about the portion *PMg* can be made equally about the portion *P'M'g*. Now since at the instant of rupture the whole mass is in limiting equilibrium, it follows that a portion *PMP'M'*, whether infinitesimal or not, acted upon by its own weight, and supported by its friction, its cohesion, and the several forces that are exerted on it by the surrounding fluid, must also be in limiting equilibrium. But it should be obvious that a mass *PMg* cannot be supported by its cohesion and its friction which prevent it slipping along *PM*, without the same friction and the same cohesion acting by reaction in the opposite sense on the mass *CBPM*. Thus denoting by *A* the horizontal force exerted on the line *PM*, and by *A'* that exerted on the line *P'M'*, any element *PMP'M'*, which must be in limiting equilibrium, will be supported along a horizontal line *Fe* by the thrust $(A' - A)$, will be loaded along the vertical line *PM* by the reactive frictional force expressed by A/n and by the reactive cohesive force δPM, and will be supported by the friction and the cohesion of *P'M'* and by the friction and cohesion of *MM'*. Thus the elementary surface *PP'MM'* could be considered equivalent to a triangle *MM'q*, loaded by the weight of the element acted upon by all the vertical forces that we have just enumerated. Let

$$gP = x \qquad PM = y$$
$$gP' = x' \qquad P'M' = y'$$
$$gP'' = x'' \qquad P''M'' = y''.$$

We have found (articles 9 *bis* and 10) that a triangular surface whose vertical side is *a* and horizontal side is *x*, loaded by a vertical force ϕ, would give the horizontal thrust

$$A = \frac{\phi(a - xn^{-1})}{x + an^{-1}} - \frac{\delta(a^2 + x^2)}{x + an^{-1}}.$$

Comparing this equation with that which would arise for the element *PP'MM'*, it will be seen that A represents $(A'-A)$, that

$$\phi = y(x'-x) + (A-A')\, n^{-1} + \delta(y-y')\ [32],$$

that

$$a = (y'-y)$$

and that

$$x = (x'-x);$$

thus the equation expressing the state of limiting equilibrium will become

$$(A'-A) = [y(x'-x) + (A-A')\, n^{-1} + \delta(y-y')]$$

$$\times \left[\frac{y'-y-(x'-x)\, n^{-1}}{x'-x+(y'-y)\, n^{-1}}\right]\ [33].$$

Assume for simplicity that $\delta = 0$, as for newly-turned soils; then

$$(A'-A) = \frac{y(x'-x)[(y'-y)-(x'-x)\, n^{-1}]}{2(y'-y)\, n^{-1}+(x'-x)\,(1-n^{-2})}.$$

By the same arguments, it will be found that

$$(A''-A') = \frac{y'(x''-x')\,[(y''-y')-(x''-x')\, n^{-1}]}{2(y''-y')\, n^{-1}+(x''-x')\,(1-n^{-2})},$$

so that, adding together these two equations,

$$(A''-A) = \frac{y(x'-x)\,[(y'-y)-(x'-x)\, n^{-1}]}{2(y'-y)\, n^{-1}+(x'-x)\,(1-n^{-2})}$$

$$+\frac{y'(x''-x')\,[(y''-y')-(x''-x')\, n^{-1}]}{2(y''-y')\, n^{-1}+(x''-x')\,(1-n^{-2})}$$

But since the horizontal thrust of the surface *PMg* must be a maximum, as must the horizontal thrust of the surface *P"M"g*, it follows that if the quantities y, y', y'' and x, x'' remain constant, then x', the only variable, must be such that $A''-A$ is a maximum. This gives, on differentiating and setting $y'-y = y''-y' = dy\ [34]$,

$$\frac{d(A''-A)}{dx'} = \frac{y(dy-(x'-x)\, n^{-1})-yn^{-1}(x'-x)}{2n^{-1}dy+(x'-x)\,(1-n^{-2})}$$

$$-\frac{(1-n^{-2})\, y(x'-x)\,[dy-n^{-1}(x'-x)]}{[2n^{-1}dy+(x'-x)\,(1-n^{-2})]^2}$$

$$-\frac{y'[dy-(x''-x')\, n^{-1}]+y'n^{-1}(x''-x)}{2n^{-1}dy+(x''-x')\,(1-n^{-2})}$$

$$+\frac{(1-n^{-2})\, y'(x''-x')\,[dy-n^{-1}(x''-x')]}{[2n^{-1}dy+(x''-x')\,(1-n^{-2})]^2}.$$

But since the different corresponding terms of this equation are consecutive similar functions, it follows that, on integrating and substituting dx for $x' - x$,

$$B = \frac{y(dy - n^{-1}dx)}{2n^{-1}dy + dx(1 - n^{-2})} - \frac{yn^{-1}dx}{2n^{-1}dy + dx(1 - n^{-2})}$$
$$- \frac{(1 - n^{-2})\, y\, dx(dy - n^{-1}dx)}{[2n^{-1}dy + dx(1 - n^{-2})]^2}.$$

If the substitution $z\, dy = dx$ is made in this equation, then

$$\frac{y(1 - zn^{-1})}{2n^{-1} + z(1 - n^{-2})} - \frac{yzn^{-1}}{2n^{-1} + z(1 - n^{-2})} - \frac{(1 - n^{-2})\, yz(1 - zn^{-1})}{[2n^{-1} + z(1 - n^{-2})]^2} = B.$$

Since z is raised only to the second power in this reduced equation, it will have the following form:

$$z^2 + \frac{F' + G'y}{F + Gy}\, z + \frac{F'' + G''y}{F + Gy} = 0,$$

and hence
$$z + \frac{F' + G'y}{2(F + Gy)} = \pm \left[\left(\frac{F' + G'y}{2(F + Gy)} \right)^2 - \frac{F'' + G''y}{F + Gy} \right]^{\frac{1}{2}},$$

where F, F' and F'' as well as G, G' and G'' are constant coefficients.

If cohesion had been taken into account, an equation of precisely the same form would have been derived, the only difference being in the coefficients.

It may be concluded from this last analysis that if a fluid of given cohesion and friction were contained within a vessel CBg', the thrust against the wall CB would be the same whatever the shape of Bg', provided that the curved surface Beg could be inscribed that would give a maximum thrust in an unbounded fluid. However, if the curve Beg which gives the greatest thrust were outside the vessel, then it would be necessary to find that surface, of all the surfaces which could be inscribed, which would give the greatest thrust.

If, however, the cohesion and the friction of the fluid with the vessel were less than the internal values of the fluid itself, then the thrust of the fluid contained in the vessel could be greater than that of the unbounded fluid [35]. The extension of these remarks, as well as the application of the above formulae, requires a special piece of work, and would take me away from the simplicity which I laid down for myself in this memoir. I hope however to be able to deal elsewhere with this subject in connection with the theory of mines, which makes use of some of the principles I have just explained, but still lacks the solution of several quite interesting problems.

XVI
On arches [36]

Let a curve *FAD* be described on the base [37] *FD*; and let a second   Fig. 9
curve *fad* be described outside the first. Let the curve *FAM* be
divided into an infinite number of portions *MM'*, and from each
point *M* let there be drawn the line *Mm* normal to the interior curve
at *M*, or making an angle with the element *MM'* according to a given
law. If it is assumed that the two lines *FAD, fad* are such that any
portion *AaMm* is in equilibrium, when acted upon by its own weight
and supported by cohesion and friction, then the elevation of an arch
will have been constructed. If now this elevation is assumed to move
parallel to itself, forming a solid envelope contained between the
surfaces generated by the two curves, then equilibrium demonstrated
for the elevation will also hold for the solid envelope. The arch formed
in this way will be that called a *voûte en berceau* [38], and it is this
type that I considered in the following investigations. The principles
developed for this arch can be applied to all other types of vaults [39].

XVII
On arches with joints that have neither friction nor cohesion

Let *aB* be the elevation of an arch of infinitesimal thickness, of which   Fig. 10
the joints are perpendicular to the curve *aB*; required the shape of
the arch when acted upon by any forces.

Let all the forces acting on the portion *aM* be resolved in two
directions, one vertical and the other horizontal; let the resultant of
all the vertical forces be *QZ*, which I will call ϕ; let the resultant of
all the horizontal forces be *Qϕ*, which I will call π; further, let
$aP = y$, $PM = x$, $Mq = \mathrm{d}y$, $qM' = \mathrm{d}x$. It is clear (article 1) that the
resultant of all the forces acting on the portion *aM* must be perpendicular to the joint at *M*, and, by article 3, that if all the forces acting
on that part of the arch are resolved in two perpendicular directions,
one vertical and the other horizontal, then the sum of the forces in
each direction must be zero. Thus if *P* denotes the thrust at the joint
at *M*, and if this thrust is resolved in two directions, one horizontal,
*P*d*x*/d*s*, and the other vertical, *P*d*y*/d*s*, then the two following
equations will be obtained:

$$P\frac{\mathrm{d}x}{\mathrm{d}s} = \pi, \quad \text{and} \quad P\frac{\mathrm{d}y}{\mathrm{d}s} = \phi.$$

Thus, on dividing one by the other to eliminate P,

$$\frac{\mathrm{d}x}{\mathrm{d}y} = \frac{\pi}{\phi},$$

an equation which expresses the shape of an arch when acted upon by any given forces.

This formula is exactly the same as that found by Euler (in the third volume of the Academy of Petersburg) for the shape of a chain acted upon by given forces. This must be so, since, on inverting the curve, and substituting tension for compression, the above theory applies equally to either case, and gives precisely the same equation. However, Euler's method has only the result in common with the method given here.

COROLLARY I

If the horizontal thrust were constant and equal to the thrust at a, and if the resultant of the vertical forces were equal to the weight of the portion aM of the arch, then

$$\frac{\mathrm{d}x}{\mathrm{d}y} = \frac{A}{\int p\,\mathrm{d}s} \quad [40],$$

from which will be found the value of p, if the curve is known, or equally the expression for the curve when the distribution of the weight p is known.

COROLLARY II

Suppose that the thickness of the arch were finite, the other assumptions being the same as in the previous corollary. Let R be the radius of curvature of the curve at the point M, and z be the length of the joint Mm; then

$$MM'mm' = \frac{z\,\mathrm{d}s(2R+z)}{2R}$$

and hence

$$\frac{\mathrm{d}x}{\mathrm{d}y} = A\left[\int \frac{z\,\mathrm{d}s(2R+z)}{2R}\right]^{-1},$$

from which

$$Ad\left(\frac{\mathrm{d}y}{\mathrm{d}x}\right) = \frac{z\,\mathrm{d}s(2R+z)}{2R}.$$

But

$$R = \frac{\mathrm{d}s^3}{\mathrm{d}(\mathrm{d}y)\,\mathrm{d}x} \quad [41],$$

and hence

$$\mathrm{d}\left(\frac{\mathrm{d}y}{\mathrm{d}x}\right) = \frac{\mathrm{d}s^3}{R\,\mathrm{d}x^2};$$

Fig. 11

thus
$$\frac{A}{R}\left(\frac{ds}{dx}\right)^2 = \frac{z(2R+z)}{2R},$$

which gives
$$R+z = \left[R^2 + 2A\left(\frac{ds}{dx}\right)^2\right]^{\frac{1}{2}}$$

as the general equation for any arch under gravity loading.

EXAMPLE

If the intrados *aMB* were a circle of unit radius, and the value of
z were required, it is clear that

$$\frac{ds}{dx} = \frac{MM'}{qM'} = \frac{1}{\cos s};$$

thus
$$1+z = \left[1 + \frac{2A}{\cos^2 s}\right]^{\frac{1}{2}}.$$

If it be assumed that at the top of the curve the length of the joint *Aa*
is *b*, then $\cos s = 1$, and $A = \frac{1}{2}(2b + b^2)$.

REMARK I

In this analysis, I have been concerned only to satisfy the first condition of equilibrium, which requires that all the forces acting on
a portion *GaMm* of the arch have their resultant perpendicular to the
joint *Mm*. However, it is easy to prove that at the same time the
second condition has been satisfied, which requires that the resultant
fall between the points *M* and *m*. Since the constant force *A* acts perpendicular to the vertical joint *Ga*, at some point *S*, it follows from
the equilibrium condition that has just been satisfied that the line of
resultants must cut all the joints at right angles, and will form a curve
parallel to the intrados *aB*. Thus for the case when the force *A* is
applied at *a*, the line of thrust would be exactly the same as *aMB*.

James Bernoulli (*Opera*, vol. 2, p. 1119), in seeking the shape of
an arch of which the voussoirs would be equal and very small, finds
two different expressions from different conditions of equilibrium;
but a mistake [42] in the angle of the tangent to the curve gave rise to
Bernoulli's error, and this has already been noted by the editors
of his Works.

REMARK II

It also follows from the general formula

$$R + z = \left[R^2 + 2A \left(\frac{\mathrm{d}s}{\mathrm{d}x} \right)^2 \right]^{\frac{1}{2}}$$

that when the arch aB forms a right angle at B with the horizontal base line EB, the length of the joint at that point becomes infinite, that is, that joint is the asymptote to the extrados GD. This follows because in the fundamental equation $\mathrm{d}s$ becomes infinite compared with $\mathrm{d}x$, and hence $R + z$ also becomes an infinite quantity. This result does not really conform with what we see done every day, since in practice the horizontal joints, instead of being infinite, are often quite small. In theory, moreover, if the curve of the intrados is given, then the length of the joint always has a precise value; this value, however, is varied widely in practice by architects. It is, however, the strength given by friction and cohesion that preserves the equilibrium that gravity tends to destroy. We will seek later the way to introduce these extra passive forces into the calculation of arches; but in the meantime it may be inferred from these comments that in practice the above theory, as we have already said in the introduction, is only of little use.

COROLLARY III

If both extrados and intrados were given, the direction of the joints for the case of equilibrium could be found in the following way.

Fig. 12   As above, assume that the joint aG is vertical, and extended indefinitely to l; let qM be the joint at M, which, extended, meets the vertical al in C; let ϕ be the centre of gravity of the portion $aGMq$; let Sp be the direction of the constant horizontal force A which meets at p the vertical passing through the centre of gravity ϕ. The resultant of all the forces will be given by a line pn, which (article 1) must be perpendicular to the joint Mq and must pass between the points M and q. Construct PM [43] parallel to the base AB, and let angle PMC be h. Since both curves aMB and GqD are given, the weight of the portion $GaMq$ will be expressed by a function of PM and of h; but the two similar triangles prn, PCM, of which the sides of the first are proportional to the forces which act on the portion $GaMq$ of the arch, give the following relationship: The weight P of the portion $GaMq$ of the arch is to A as is $\cos h$ to $\sin h$, that is $P = A \cos h / \sin h$. We will see later which are the points S between a and G at which the

64

thrust A, determined from the previous expression, can be applied so as to satisfy the second equilibrium condition, namely that the resultant pn always pass between the points M and q.

EXAMPLE

Suppose the direction of the joints were required for a plate-bande of constant given thickness; let $aGBb$ represent this arch contained between two parallel straight lines. The directions of the vertical joint aG and of the last joint Bb, by which the arch is supported on the wall $BLKo$, being given, the direction of all the other joints Mm is required. Let $aG = a$, $aM = x$, and let the direction of the joint Mm meet the vertical aG in C; then

$$GaMm = P = ax + \frac{a^2 \cos h}{2 \sin h}.$$

Substituting this value of P in the fundamental equation

$$A \frac{\cos h}{\sin h} = P,$$

then
$$ax = \left(A - \frac{a^2}{2} \right) \frac{\cos h}{\sin h}.$$

To determine the value of the constant A, suppose that when

$$x = aB = b, \quad \frac{\cos h}{\sin h} \text{ equals } C.$$

It will be found that

$$A = \frac{2ab + a^2 C}{2C}, \quad \text{and hence} \quad x = \frac{b \cos h}{C \sin h};$$

from which it may be deduced that all the joints of a plate-bande pass through the same point C, which makes construction very easy.

The satisfaction of the second condition of article 1 requires in this example that the resultant of the forces which keep the portion $GaMm$ of the arch in equilibrium should pass between the points M and m. Let ϕr be the vertical line passing through the centre of gravity of the whole mass $GaBb$. If on the joint Bb is erected at the point B a perpendicular Bn, which meets the vertical ϕr at n [*44*], and if through this point n is drawn the horizontal line ns, then the point s where the vertical joint Ga will be met by this line will be the lowest point on the joint Ga at which the force A can be applied without the plate-bande collapsing. Thus if the direction of the joint Bb were

Fig. 13

65

such that the line *Bn* would meet the vertical ϕr in a point *n* above the line *Gb*, there would be no point on the joint *Ga* at which the force *A* could be applied to maintain equilibrium, and the plate-bande would inevitably collapse. From these remarks, it is very easy to determine the limit of the slope of *Bb* when the thickness *Ga* is given.

It should not be necessary to point out that if the resultant *Bn* for the whole mass passes through *B*, then the resultant for a particular mass *GaMm* will certainly pass between *M* and *m*; since, while the quantity *A* remains fixed, the masses *GaMm* decrease. Thus, once the second equilibrium condition has been satisfied for the point *B*, the same condition will necessarily have been satisfied for any point *M*.

XVIII

On the stability of arches, allowing for friction and cohesion

PROBLEM

Fig. 14 *The intrados aB and the extrados Gb are given for an arch, as are the joints Mm normal to the intrados; required the limits of the horizontal thrust at f which would support the arch, when it is acted upon by its own weight and sustained by the cohesion and the friction of the joints.*

Take a portion of the arch, such as *GaMm*. Extend *mM* to *R* and denote the angle at *R* by *h* [45]; let the thrust applied at *f* to the vertical joint *aG* be denoted by *A*.

I assume first that the portion *GaMm* is solid, so that it could break only along *Mm*. Thus, if this portion of the arch is to be stable, the thrust *A* must be such as to prevent it sliding along *mM*; but

the resolved part of the force *A* along *Mm* is: $\quad A \sin h.$

The force parallel to *mM*, component of ϕ, is: $\quad \phi \cos h.$

The force perpendicular to *mM*, component of *A*, is: $A \cos h.$

The force perpendicular to *mM*, component of ϕ, is: $\phi \sin h.$

Thus, allowing for friction and cohesion, the tendency for the portion of the arch to slide along *mM* will be expressed by

$$\phi \cos h - A \sin h - \frac{\phi \sin h + A \cos h}{n} - \delta . Mm;$$

and in the case when *A* will be only just suffiicent to support it, then

$$A = \frac{\phi(\cos h - n^{-1}\sin h) - \delta . Mm}{\sin h + n^{-1}\cos h}.$$

Now since the actual construction will allow the arch to slide not only on the joint mM, but equally on all others, it follows that if the arch is not to collapse, A must never be less than the quantity

$$\frac{\phi(\cos h - n^{-1}\sin h) - \delta . Mm}{\sin h + n^{-1}\cos h},$$

whatever the value of h. Thus if the value of h is taken as that which gives a maximum for A, then the value of A found in this way will be sufficient to support the whole arch.

I take A_1 to denote this maximum.

If the force at f were required which would just cause the portion of the arch offering the least strength to slide along Mm, then for any portion at the state of limiting equilibrium,

$$A = \frac{\phi(\cos h + n^{-1}\sin h) + \delta . Mm}{\sin h - n^{-1}\cos h};$$

but since no portion of the arch must slide along any joint Mm, then A must always be less than this last expression. Thus the minimum value of A must be sought, which will then give the largest force that can be applied at f without breaking the arch along the joint Mm; I take A' to be this minimum.

Now since, for equilibrium, which is the case we wish to determine, the arch, in whole or in part, must not slide at its joints in either direction, it follows that the limits of the forces that can be applied at f lie between A_1 and A', where A_1 represents the least thrust that can act at f, and A' the greatest thrust that can act at the same point. Thus it may be concluded that if A_1 is greater than A', equilibrium cannot be achieved, since the thrust at f cannot be greater than A', nor can it be smaller than A_1, which we suppose to be greater than A'.

In order to satisfy the second condition of equilibrium, the resultant of all the forces acting on the portion $GaMm$ of the arch must pass above the point M and below the point m [46]. Thus, denoting by B the force which acts at f, $B . MQ$ must always equal or be greater than $\phi . gM - \delta' . z^2$ (where δ' [47] is a constant fraction of the cohesion of the mortar, article 7), and in the case when the resultant passes through M, then

$$B = \frac{\phi . gM - \delta' . z^2}{MQ}.$$

If the quantity B were supposed smaller than $(\phi . gM - \delta' . z^2)/MQ$, then the resultant would pass below the point M [48], and the arch would collapse. Thus, to find the force B sufficient to hold up the whole arch, the maximum value of B must be sought from the above equation, and this maximum will give the smallest force which can act at f; let B_1 denote this maximum.

Further, to satisfy the second condition, the resultant must pass below the point m, so that it follows that $B.mq$ must be smaller than, or at most equal to $\phi.g'q + \delta'.z^2$. Thus the constant B must be determined from the equation

$$B = \frac{\phi.g'q + \delta'.z^2}{mq}$$

so that it represents the minimum of $(\phi.g'q + \delta'.z^2)/mq$. The value B' found in this way will make the quantity $B.mq$ equal to $\phi.g'q + \delta'.z^2$ at one point only, and smaller at all the other points m, and hence B' will express the greatest force that can be supposed to act at f. Thus it may be concluded that, to satisfy the second condition, the force applied at f cannot be smaller than B_1 nor greater than B'. Hence, putting together the two conditions, if either A_1 or B_1 is greater than A' or B', equilibrium is not possible, and the arch of given dimensions would necessarily collapse.

To establish in fact the true limits, it is sufficient only to choose from A_1 and B_1 the larger, and from A' and B' the smaller, so that if B_1 were larger than A_1 and B' smaller than A', then B_1 and B' would be the true limits of the forces which could be applied at f without breaking the arch.

REMARK I

Friction is often large enough in the materials used for arch construction that the different voussoirs could not slide one on another. In this case, the first equilibrium condition can be ignored, and it is no longer necessary that the resultant of the forces which act on any portion of an arch should be perpendicular to the end joints of that portion, but only that it should fall within those joints. Thus, neglecting the cohesion of the joints, which should be done for newly-built arches, it is sufficient to find the maximum of $\phi.gM/MQ$ to determine the force B_1, and the minimum of $\phi.qg'/mq$ to determine B'; further, the force B must be assumed to act at G, the top of the joint, in order to make the force B_1 as small as possible. However, it must be noted that when stability is established by means of this second condition, by supposing that the thrusts pass through the points G and M, it must be assumed that these points are far enough away from the extremities of the joints so that the cohesion of the voussoirs prevents the thrusts from crushing the edges [49]. This can be determined by the methods we used in finding the strength of a pier.

REMARK II

In practice it will always be easier to find the limits of the thrust B by trial and error rather than by exact methods. Assume, for example, that a portion GaM of the arch is taken such that the joint Mm makes an angle of $45°$ with the horizontal; then the force B_1 will be calculated on this assumption. This same force will then be calculated with respect to a second joint, near the first and towards the keystone. If this second force is greater than the first, it will be certain that the rupture point of the arch is between the keystone and the first joint; thus moving towards the keystone and repeating the operation, the actual force B_1 will easily be found. This calculation would never be very lengthy, because by the properties of maximum and minimum there will be, near the point M where the required limit B_1 is found, very little variation over quite a large portion of the curve. Thus, to determine the force B_1, the rupture point M need be known only approximately. The greatest force B' that an arch could carry without breaking will be found by the same method. Hence if the dimensions of an arch are given, as we assume here, and also the height of the pier BE on which it bears, the necessary thickness Bb of the pier will easily be found such that the resultant of the thrust B_1 acting at G and of the total weight of the arch [50] and of its pier should pass between E and e, or should pass through e; this will satisfy the second condition of stability.

This Mémoire is perhaps already too long, and its scope does not allow me to extend this theory, nor to apply it to all kinds of arches; thus I will content myself with having tried to give exact methods, and those that I consider essential to establish the state of stability.

Comparing the above principles with the different approximate methods used in practice, it will easily be seen that their authors were not sufficiently aware of the two equilibrium conditions necessary for the state of stability. In the method, for example, attributed to La Hire, as reported by Bélidor, and used by almost all builders, the arch is divided into three portions, and the thrust of the upper part is calculated in accordance with the first equilibrium condition; the dimensions of the piers are then found by the second equilibrium condition. Now it should be obvious that if the upper portion is divided near the keystone, and if it is supposed that the arch breaks in four portions, rather than three, the thrust of the upper portions will often be, for flat arches, much greater than that found by the method of La Hire, and the size of the piers calculated by his method will often be inadequate [51].

NOTES

[*1*] Three of the many meanings given by Littré for the word *essai* are:

(*a*) the first attempts of someone working in a given field,

(*b*) a work in which the author treats the subject without pretending to say the last word,

(*c*) the title of many specialist works, given out of a feeling of modesty, as if the word treatise might be too weighty.

Each of these three meanings fits Coulomb's *Mémoire*; the feeling of modesty is perhaps excessively prominent at the end of the introduction (p. 4; p. 43).

Littré would really wish that *maximum* were treated as a French word, but concedes that mathematicians use the ablative plural in such phrases as 'Méthode de maximis et minimis'.

The word *architecture* could, in this context, be translated by the phrase 'civil engineering'.

[*2*] Kerisel draws attention to Coulomb's use of the word *couler*, and calls him 'le devancier des plasticiens'; certainly the search among possible surfaces of failure to find that which gives the least value of collapse load is a technique of modern plastic theory. However, Coulomb was in fact using standard words of his time; in particular, *couler* occurs repeatedly in Bullet's *Architecture pratique* of 1691, a work which may well be regarded as giving the first treatment of the mechanics of soils.

[*3*] It is clear that Coulomb is discussing the *limiting* state.

[*4*] The words *cohésion*, *cohérence* and *adhérence* are used almost interchangeably. In the translation, cohesion will usually refer to stresses, and coherence to cohesive *forces*.

[*5*] Assuming that simple plastic theory may be applied to this problem, the technique will lead to upper bound solutions. For a postulated mechanism of collapse (i.e. one portion of the pier sliding on the other), calculation will lead to an *unsafe* estimate of the fracture load. Thus, for any assumed inclined plane, the resulting value of the load will be too large; that plane will be correct (providing that fracture does indeed occur on a plane surface) which gives a minimum value of the load. This minimum load is the maximum load the pier can carry; the value for any other assumed mechanism is greater than the minimum.

[*6*] The soil is considered here to be in an *active* state. The assumed mechanism again leads to an unsafe estimate of the value of the force necessary to prevent the soil slipping *down*; that is, the largest value of that force must be found by consideration of different rupture surfaces. Coulomb envisages the possibility of a curved rupture surface (see article 15 of the *Mémoire*), but his main calculations are concerned with the rectilinear wedge. Article 10 considers the *passive* problem of finding the largest horizontal force that can be applied without the soil slipping *up*. Again, the assumed mechanism leads to a value of the force which must be minimized, exactly as for the pier.

[*7*] This paper is considered below, pp. 75ff, together with the other works referred to by Coulomb. Actually Hooke had formally identified the problems of the arch and of the catenary in 1675, some twenty years before Gregory.

[*8*] This innocent proposition contains the germ of much of the subsequent work

70

in the *Mémoire*. A body resting on a smooth horizontal plane will be in equilibrium if the resultant of all the forces acting on the body is perpendicular to the plane, and falls within the base area of the body. There is no attempt to define the *actual* resultant of the forces, either in magnitude or position; if the resultant can be shown to satisfy the stated conditions, then equilibrium is assured, and there is no need to find the actual line of action. Further, there is no need to find the magnitude of the resultant (although this may in fact perhaps be found quite easily). The proposition is thus not numerical but geometrical; an equilibrium statement is replaced by a geometrical statement.

[9] That is, the coefficient is $\frac{3}{4}$.

[10] The three conditions must hold whatever the stress–strain law of the material. The stress distribution of fig. 6 is indeed shown as a general non-linear distribution.

[11] That the line *fh* remains straight after straining is simply stated by Coulomb; it is in fact an assumption, namely that plane sections remain plane.

[12] The effect of small shear stresses is ignored.

[13] Coulomb is finding the ultimate moment of resistance of a material with a constant finite yield stress in tension and infinite compressive strength. The 'cohesive triangle' should really be a 'cohesive rectangle', but the analysis is correct, and is really concerned with the determination of the ultimate moment of resistance of a member whose tensile and compressive yield strengths differ. Coulomb discusses in the next paragraph the corrections to be made to allow for a finite rather than an infinite compressive yield strength.

[14] The beam has been 'cut' on the (vertical) line *fh*, and the section *fh'* is subject to a uniform tensile stress δ. Thus an equal and opposite compressive force (of magnitude $\delta . fh'$) must act on *hh'*. This compressive force will cause failure of the masonry along some line *h'q*, which can be determined using the methods of article 8.

[15] In fact, Coulomb uses only the shear strength δ, and does not require the value of the tensile strength in his analysis; he merely quotes the tensile strength on p. 13; p. 50, and again in analysing Musschenbroek's tests.

[16] Page 359 in the 1739 edition. The tests are reported in the course of chapter 19, *De l'Adhérence des Corps*, pp. 344–61.

[17] The *ligne* was the twelfth part of an inch. Littré gives its value as 2·1166 mm, but Larousse gives 2·25 mm, which agrees with Kerisel's statement that at the time of Vauban the foot was 0·3248 m. Thus Coulomb's *pied* is about 12·8 in, and his *pouce* about 1·06 in.

[18] In article 6 Coulomb actually stated that the tensile strength of brick lay between 280 and 300 lb/in². Using the smaller value, Musschenbroek's compression test should have given 194 lb.

[19] There are several misprints in the original equations, and these will be corrected, as here, usually without comment; as noted by Gillmor, Coulomb complained in 1781 that he had not been able to proof-read his *Mémoires*. In writing this equation Coulomb almost, but not quite, succeeded in setting up general expressions for the forces acting on a triangular element of a continuous medium. Had he done so explicitly, the equations of equilibrium of stresses, and indeed the whole development of continuum mechanics, might have become established before Cauchy. However, Coulomb did succeed in showing, p. 11; p. 48, that the

plane of maximum shearing stress made an angle of 45° with the direction of a (single) principal stress.

If the triangle *CaB* is isolated, then it is immediately clear that

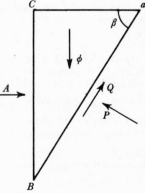

$$P = A\sin\beta + \phi\cos\beta, \\ Q = -A\cos\beta + \phi\sin\beta.$$

Coulomb's yield criterion is $Q = Pn^{-1} + \delta \cdot aB$ (but is nowhere stated by him in this explicit form), and his result is obtained on substituting in the values of P and Q.

[20] It will be seen that the maximum of the expression for A (and, similarly, the minimum of that for A') will occur for $\delta = 0$. It is then easy to show that n^{-1} should also be zero. Some of the working is implicit at the end of Coulomb's article 10, below.

[21] It may be noted that not only are the values of the forces A and A' different for the active and passive cases, but so also are the rupture lines.

[22] Here, and in later working, Coulomb sometimes omits the unit soil weight g. The basic results for angle of slip are of course independent of g; later on, when calculating the dimensions of retaining walls, a numerical example is worked on the assumption that the density of the masonry is the same as that of the soil, so that again g can be omitted.

[23] $m = \frac{1}{2}g[(1+n^{-2})^{\frac{1}{2}} - n^{-1}]^2$, $l = 2[(1+n^{-2})^{\frac{1}{2}} - n^{-1}]$.

[24] Thus the *whole mass* of soil behind the retaining wall is assumed to be in a state of limiting equilibrium, and to be slipping along rupture planes parallel to *Ba*.

[25] In fact, $x = (\sqrt{2}-1)\,a$, $A = (\frac{3}{2} - \sqrt{2})\,ga^2 = 0.086ga^2$, which is the value given by Coulomb. The unit weight g is omitted in the original, since the retaining wall is taken as having the same density as the soil.

[26] Following through the 'exact' calculations, $c = (1 - \frac{2}{3}\sqrt{2})^{\frac{1}{2}}\,b = 0.24b$.

[27] That is, the thrust A is less than the maximum horizontal friction force that can be developed at the base *GE*.

[28] The misprint of *GE* for *CE* is corrected in the 1821 edition. Coulomb's analysis of article 14 below shows that the value of A is diminished by the effect of friction between the soil and the retaining wall; the vertical friction force along *CE* will oppose the overturning of the wall.

[29] Compare the well-known result for the limiting slope β of a saturated cohesionless soil of internal friction angle ϕ:

$$\tan\beta = \left(1 - \frac{\gamma_\omega}{\gamma}\right)\tan\phi.$$

Here γ_ω and γ are the density of water and the bulk density of the soil. For $\phi = 45°$ and $\gamma_\omega/\gamma = \frac{1}{2}$, $\beta = 26°\,34'$, which agrees with Coulomb's 60° to 70°.

[30] This article is not easy to follow, and Coulomb makes a mistake (see next note) which leads him to the wrong result (foot of p. 27; p. 60). In fact, the solution of the basic Coulomb problem, that of thrust of soil, whether cohesionless or not, against a frictionless vertical retaining wall, involves slip along straight

characteristics. Thus the surface of greatest thrust, *CBg* in fig. 8, is the triangular prism *CBa* of fig. 7.

[*31*] Coulomb has already assumed that slip is occurring along *gM*, and limiting stresses have been specified as acting on this line. It is not then possible, in general, also to specify limiting stresses on the vertical line *PM*. Again, what was lacking (see note 19 above) was a full analysis of equilibrium of a continuous medium.

[*32*] As usual, the soil density is taken as unity.

[*33*] A term in δ is already missing from this equation.

[*34*] In fig. 8, the verticals *PM* and *P"M"* are fixed, and *P'M'* is varied to maximize $(A''-A)$; this leads to the displayed equation on p. 27; p. 59, for which all the terms on the right-hand side must sum to zero. The first two terms refer to the slice *PMP'M'*, and the second two to *P'M'P"M"*, and are 'consecutive similar functions'. Since the position of the slices is arbitrary, it follows that each group of two terms for any one slice must sum to a constant. This constant is denoted *B* in the next equation, which is written for the limiting case of an infinitesimal slice.

[*35*] It is indeed 'obvious' that a reduction in the value of coefficient of friction between the soil and retaining wall must lead in some sense to a 'worse' case. That this statement is not necessarily true was demonstrated by Drucker (1954).

[*36*] *Voûte* means, more generally, vault, but Coulomb deals with two-dimensional arches only, as the definition in article 16 makes clear. While the word *arche* is used, as for example in *arches du pont*, nevertheless the form of those same bridge arches will be described as *voûte en berceau* etc. (see below).

[*37*] The 'axis' of an arch in Coulomb's, as in modern French, usage, means the line joining the springings; the arch is symmetrical about the perpendicular bisector of this 'axis'.

[*38*] The *voûte en berceau* is, as noted by Coulomb, any vault whose intrados is cylindrical, i.e. one whose profile is constant. If the intrados is the arc of a circle, then the arch is a *voûte en arc de cercle* (or, more shortly, *voûte en arc*), and if the intrados is a complete semicircle, then the arch is said to be *en plein cintre*. None of the arches sketched by Coulomb in figs. 9, 11, 12 and 14 has a circular intrados, but all are of the oval type called *voûte en anse de panier* (basket-handled arch). These three- or four-centred (in general, many-centred) arches all rise vertically from their springings (as shown in fig. 14); the *voûte en ellipse* is a special case of this general type. As Coulomb notes immediately, his theory is not restricted to any particular profile (see note 39 below).

[*39*] It is true that the *general* principles developed by Coulomb can be applied to a three-dimensional vault, but the application is somewhat removed from the specific calculations made here. It is likely that the remark imports merely that the principles can be applied to an arch of any profile.

[*40*] The constant horizontal thrust is *A*, and *p* is the weight per unit length of arch.

[*41*] A unit weight of material is assumed. In more modern notation

$$A \frac{d^2y}{dx^2} = z \left(\frac{ds}{dx}\right) \frac{(2R+z)}{2R}, \quad R = \left(\frac{ds}{dx}\right)^3 \Big/ \frac{d^2y}{dx^2}.$$

[*42*] In dealing with an infinitesimal segment of the arch, Bernoulli is wrong by a factor 2 in the value of a small change of angle. This is noted on p. 1120 of the 1744 edition of the *Opera*.

73

[*43*] The point *P* is shown as *F* in fig. 12. The letters *LKo* are missing from fig. 13 below; see note 44.

[*44*] Figure 13 has been redrawn as shown.

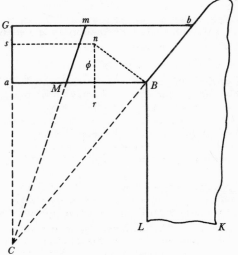

[*45*] This angle *h* is the complement of the angle *h* in the previous article.

[*46*] See note 48 below.

[*47*] The distance *Mm* in fig. 14 is equal to *z*. From article 7, p. 9; p. 47, the quantity δ' is either $\frac{1}{8}\delta$ or $\frac{1}{2}\delta$, depending on whether the 'elastic' or the 'rigid' hypothesis is made. Coulomb leaves this question open, but in any case soon makes the 'safe' assumption (Remark I, p. 38; p. 68) that the cohesion δ should be neglected.

[*48*] If the moment of resistance of the mortar, $\delta'z^2$, is taken into account, then the resultant *can* fall below *M*. The limiting case of passing through *M* occurs, as noted by Coulomb in his Remark I, p. 38; p. 68, when $\delta' = 0$.

[*49*] The general stress level in masonry in usually extremely low, so that thrust lines can approach close to the surface without crushing the stone.

[*50*] The weight of half the arch.

[*51*] Coulomb notes here that the method of La Hire gives reasonable assurance that a satisfactory thrust line can be found for the lower portion of the arch and the piers. No calculations are made, however, to ensure that the thrust line remains within the masonry for the upper portion of the arch, and this condition may indeed be violated for arches with flat crowns, which would lead to an unsafe design.

2

Coulomb's References

DAVIDIS GREGORII, Catenaria, *Philosophical Transactions* no. 231, 637 (1697).

GUILLAUME AMONTONS, De la résistance causée dans les machines, tant par les frottemens des parties qui les composent, que par la roideur des cordes qu'on y employe, et la maniere de calculer l'un et l'autre, *Histoire de l'Académie Royale des Sciences 1699*, 206, Paris (1702).

CHARLES BOSSUT and GUILLAUME VIALLET, *Recherches sur la construction la plus avantageuse des digues*, Paris (1764).

PIERRE VAN MUSSCHENBROEK, *Essai de Physique*, translated from the Dutch by Pierre Massuet, Leyden (1739).

LEONHARDO EULERO, Solutio problematis de invenienda curva, quam format lamina utcunque elastica in singulis punctis a potentiis quibuscunque sollicitata, *Commentarii Academiae Scientiarum Imperialis Petropolitanae 1728*, **3**, 70, Petersburg (1732).

JAMES BERNOULLI, *Opera* (2 vols.), Geneva (1744).

PHILIPPE DE LA HIRE, *Traité de Mécanique*, Paris (1695).

PHILIPPE DE LA HIRE, Sur la construction des voûtes dans les édifices, *Mémoires de l'Académie Royale des Sciences 1712*, 69, Paris (1731).

BERNARD FOREST DE BÉLIDOR, *La science des ingénieurs dans la conduite des travaux de fortification et d'architecture civile*, Paris (1729).

SÉBASTIEN LE PRESTRE DE VAUBAN, *Traité de l'attaque des places*, Paris (1704); *Traité de la défense des places*, Paris (1706).

DAVID GREGORY (1661–1708). The paper 'Catenaria' applies Newton's method of fluxions to determine the shape of a hanging chain. The properties of the catenary are discussed, but it is Gregory's corollaries which are of present interest. The following translation of part of corollary 6 is due to Ware (1809):

> In a vertical plane, but in an inverted situation, the chain will preserve its figure without falling, and therefore will constitute a very thin arch, or fornix; that is, infinitely small rigid and polished spheres disposed in an inverted arch of

a catenaria will form an arch; no part of which will be thrust outwards or inwards by other parts, but, the lowest part remaining firm, it will support itself by means of its figure. For since the situation of the points of the catenaria is the same, and the inclination of the parts to the horizon, whether in the situation *ACB*, or in an inverted situation, so that the curve may be in a plane which is perpendicular to the horizon, it is plain it must keep its figure unchanged in one situation as in the other. And, on the contrary, none but the catenaria is the figure of a true legitimate arch, or fornix. *And when an arch of any other figure is supported, it is because in its thickness some catenaria is included.* Neither would it be sustained if it were very thin, and composed of slippery parts. From Corol. 5 it may be collected, by what force an arch, or buttress, presses a wall outwardly, to which it is applied; for this is the same with that part of the force sustaining the chain, which draws according to a horizontal direction. For the force, which in the chain draws inwards, in an arch equal to the chain drives outwards. All other circumstances, concerning the strength of walls to which arches are applied, may be geometrically determined from this theory, which are the chief things in the construction of edifices.

Ware objects to the whole of this passage, and produces specious arguments to try and demolish its validity. Ware himself, however, added the italics (*Et cujuscunque alterius figurae Arcus ideo sustinetur, quod in illius crassitie quaedam Catenaria inclusa sit*), and this is a very powerful statement. Translating into modern terms, and broadening, Gregory asserts that if any thrust line can be found lying within the masonry, then the arch will stand.

This idea, together with that of the polished spheres of the inverted catenary, were used brilliantly by Poleni (1748) in his demonstration of the stability of the dome of St Peter's. As has been mentioned (note 7), Robert Hooke anticipated Gregory in 1675. To his book on helioscopes is added, 'to fill the vacancy of the ensuing page', a series of anagrams together with brief descriptions of the subjects to which they relate, among them (no. 3) the famous *ut tensio sic vis*. No. 2 concerns 'the true Mathematical and Mechanical form of all manner of Arches for Building', and the anagram yields 'Ut pendet continuum flexile, sic stabit contiguum rigidum inversum'; Truesdell gives 'As hangs the flexible line, so but inverted will stand the rigid arch.'

The last sentence quoted above of Gregory's corollary insists that the question of stability of arches and walls reduces to the solution of a mathematical problem.

GUILLAUME AMONTONS (1663–1705) had previously put forward the idea that the frictional force between two bodies depended only on the force pressing them together, and not on the area of contact. As is noted in the *Histoire* of the 1699 volume of the *Académie*: 'Cette nouveauté causa quelque étonnement à l'Académie. M. de la Hire consulta aussi-tôt l'experience.' (The *Mémoires* of the *Académie* were always published at this time with a preface consisting of critical summaries; in the 1699 volume, for example, the *Histoire* occupies 123 pages, and the *Mémoires* occupy 284 pages.) M. de La Hire is the same La Hire who writes on masonry arches, and whose work is discussed below. In this case he made experiments on frictional forces and confirmed the hypothesis of Amontons. The *Histoire* concedes that Amontons has priority on the subject, and his *Mémoire* gives experimental results of friction between solid bodies, and for ropes passing round cylinders. For this last problem Amontons publishes lengthy tables giving the frictional force for ropes of from $\frac{1}{12}$ in to $2\frac{1}{2}$ in diameter loaded from 1 to 100000 lb.

Amontons concluded from the tests on solid bodies that the frictional force did indeed depend only on compressive force, that it was the same for iron, copper, lead and wood, if the contact surfaces were lubricated, and that in these cases the coefficient of friction was $\frac{1}{3}$.

CHARLES BOSSUT (1730–1814). There is no reason to doubt the sincerity of Coulomb's very warm tribute to Bossut (p. 10; p. 47). Bossut had taught Coulomb at the engineering school at Mézières, and more will be said of this, and of the later close professional relationship between the two, in chapter 7 below.

The essay on dykes won the *Prix quadruple* proposed by the Royal Academy of Toulouse in 1762; it was published in Paris in 1764 as an 'ouvrage pour servir de suite à la seconde Partie de l'Architecture Hydraulique de M. Bélidor', and deals with several problems, including those of the design of quay walls, jetties, weirs, wharves, and caissons, as well as that of the best shape of spillway from a reservoir. In discussing the failure of a retaining wall, summarized below, Bossut says that

the coherence (*adherence*) between two surfaces arises from the interlocking (*engrenement*) of their parts. This force is analogous to the resistance that a wooden beam, fixed in a wall and loaded with a weight, opposes to its fracture; but it must be noted that between these two kinds of forces there is the difference that the fibres of a wooden beam are flexible and extensible, which means that the resistance offered is not the same over the whole section on which fracture occurs, whereas the coherence between two neighbouring surfaces of the dyke, since it arises from the interlocking of hard portions lacking all elasticity, must be the same over the whole section.

This appears to be the difference between the behaviour of wood and stone noted by Coulomb on p. 10; p. 47; wood should be treated as elastic at fracture, and stone (or earth) as 'rigid'.

In his design of a simple retaining wall, Bossut makes two alternative design assumptions. Both hypotheses refer to limiting states of equilibrium; in the first, the wall is assumed to remain solid, and fails by overturning about its toe. In the second, which Bossut considers more appropriate for earth dykes rather than masonry, it is assumed that each horizontal section is on the point of failure, that is, the dyke 'tend à se diviser par tranches horizontales'.

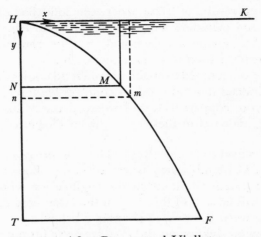

Fig. 2.1. After Bossut and Viallet.

Taking the special case of a dyke with the downstream face *HN* vertical, and for which the water level comes just to the top of the dyke, fig. 2.1, Bossut finds the shape of the curved upstream face *HM*. He assumes that failure at the level *MN* occurs by rotation about *N*, so that separation occurs along *NM*. The overturning moment due to the water pressure will thus be resisted not only by the weight of the portion *HNM* of the dyke, but also by the cohesion of the material, taken to be uniform over the length *NM* (i.e. Bossut makes the 'rigid' assumption, although at the end of his analysis he states that the value of the cohesion is to be found by tests). If p is the unit weight of water and π the unit weight of the material of the dyke, and denoting the cohesive force per unit length by πN, Bossut obtains easily the equilibrium equation for *HNM*:

$$p\frac{y^3}{6} = \int pxy\,\mathrm{d}x + \int \frac{\pi x^2}{2}\,\mathrm{d}y + \frac{\pi N}{2}x^2. \tag{2.1}$$

Since by hypothesis the equation must hold for all sections NM, it may be differentiated to give

$$\tfrac{1}{2}(ny^2 - x^2)\frac{dy}{dx} = nxy + Nx, \tag{2.2}$$

where $n = p/\pi$. This equation integrates simply by making the substitution $Z = 2(N + ny)$.

If the cohesion is neglected, i.e. $N = 0$, equation (2.2) becomes homogeneous, and gives the straight-line solution

$$y = \left(2 + \frac{1}{n}\right)^{\frac{1}{2}} x. \tag{2.3}$$

Bossut notes that the neglect of the cohesion can only improve the safety of the dyke ('ne fait d'ailleurs que concourir à la solidité de la digue'). Taking the ratio of unit weights $n = p/\pi$ as $7/10$, equation

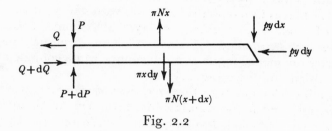

Fig. 2.2

(2.3) gives $x/y = 13/24$, almost exactly. Thus the practical rule of a batter of 1 in 2 accords well with this theory, particularly when the upstream surface is faced with masonry rubble. Bossut notes further that the downstream surface HT cannot be maintained vertical unless it too is faced with masonry; otherwise, the surface must be given a slope 'suivant le degré de fluidité des terres'.

The division of the dyke into slices, as in fig. 2.1, and the writing of an equilibrium equation for a finite portion HNM, is a method of attack closely followed by Coulomb in his article 15, pp. 24ff; pp. 57ff. Equation (2.2) can in fact be derived directly by considering the equilibrium of a thin slice of the dyke, fig. 2.2.

Bossut remarks (on p. 10) on the existence of what would now be called pore-water pressures:

It is clear that in the case that the earth in front of the foundation of the upstream face were not perfectly contiguous with this face, the water would penetrate this void ('s'insinueroit dans ce vuide') and would thrust on the dyke in this place with a head equal to the height of the water level above the bottom of the undermined foundation.

PIERRE (PETRUS) VAN MUSSCHENBROEK (1692–1761). In 1729 Musschenbroek published in Leyden his *Physicae experimentales, et geometricae...Dissertationes*; this book contains the long section (pp. 421–672) 'Introductio ad cohaerentiam corporum firmorum'. Musschenbroek gives the results of a large number of material tests in tension, bending, and compression, mainly on a variety of types of wood; some tensile tests on metal wires are also reported. Descriptions and engravings of the testing machines are included (Timoshenko reproduces some of these).

The bending tests confirm Galileo's result that the ultimate strength of rectangular beams is proportional to the width and to the square of the depth. The compression tests are the first recorded for struts, and Musschenbroek derived experimentally Euler's later result that the buckling load was inversely proportional to the square of the length. Thus Musschenbroek was dealing with slender wooden columns, whereas Coulomb's concern was with stocky masonry piers.

The 'Introductio...' starts (p. 431): 'Vocamus *Cohaerentiam, Firmitatem* vel *Resistentiam Solidorum* eam vim corporum majorum, qua partes quomodocunque et â quacunque causâ conjunctae resistunt divulsioni aut fracturae...'

Coulomb's reference is to the 1739 *Essai de Physique*, which contains the note that it was translated from the Dutch. Chapter 19 of this work, *De l'Adhérence des Corps*, contained in pp. 344–61, starts: 'Nous appellons *Adhérence*, ou *Cohésion*, cette condition et force des Corps, par laquelle leurs parties s'opposent à leur séparation...', which is an exact translation of the Latin of 10 years earlier. Chapter 19 of the French text is, in fact, a summary in under 20 pages of the 250 pages of the Latin; strengths of various materials are recorded, but there is no description of test apparatus. However, some experimental results are added in the French; in particular, the test on a brick pier does not appeal in the Latin. Only a few such 'stocky' tests are quoted, and Musschenbroek does not modify his 'Euler formula'.

The brick column broke at 195 lb (Coulomb p. 13; p. 50); the Euler buckling load (assuming an elastic modulus of 3×10^6 lb/in^2) is about 560 lb. A test on a brick column of twice the length might thus have given a breaking load of something under 140 lb. It is perhaps significant that, in refuting Musschenbroek, Coulomb chooses to halve the length rather than double it.

Stress levels, both in timber construction and in masonry, are extraordinarily low (Heyman 1967, 1966), and this is implicit in

Coulomb's 'skyscraper' fantasy (p. 14; p. 50); in practice, there is relatively little danger of failure either by instability or by crushing. Coulomb himself considers neither mode in his analysis of arches, and Musschenbroek's results on material strength and stability of compression members are not relevant to arch design.

Musschenbroek mentions in passing the problem of piers in churches, and computes a breaking load; even for piers supporting a superstructure, however, stresses remain very small. This wider application is reflected in the short paragraph in Coulomb's p. 2; p. 42, and later in pp. 14–15; pp. 50–51.

It will be noted that Musschenbroek uses *adhérence* and *cohésion* as synonyms.

LEONHARD EULER (1707–83). The paper cited is an early work, published 16 years before the book *Methodus inveniendi lineas curvas maximi minimive proprietate gaudentes,* which applies variational methods to the problem of determining the shape of a hanging chain and other problems, including the famous one of the buckling of columns. Euler returned to this last problem in 1757 in his paper: 'Sur la force des colonnes', *Mém. Acad. Berlin.*

The 1728 paper refers to both James and Daniel Bernoulli, and is a correct and dull catalogue of the shape of a flexible chain under various loading conditions. Coulomb notes justly, p. 30; p. 62, that although both he and Euler have the same general formula, there is nothing further in common. Euler deals with a flexible chain whose shape follows exactly the line of tension, while Coulomb is dealing with an arch of finite thickness whose thrust line is not constrained to follow the shape of the centre line of the arch.

JAMES (JACOB, JACQUES, THE ELDER) BERNOULLI (1654–1705), elder brother of John Bernoulli, uncle of Daniel and Nicholas; Jacques (the younger) (1759–89), was a nephew of Daniel.

The paper 'Problema de Curvatura fornicis, cujus partes se mutuo proprio pondere suffilciunt sine opere caementi', *Opera,* vol. 2, p. 1119, is included among the *Varia Posthuma.* The paper is of only five pages, and contains little that was not given by Gregory. Bernoulli's first (correct) solution is noted by him to be identical with that of Gregory ('...quod indicat curvam Catenariam, ut habet GREGORIUS.'). His second solution is, as stated by Coulomb, wrong, and there seems little point in referring to Bernoulli at all.

However, it is of interest that Coulomb knew Bernoulli's work. In the same vol. 2 of the *Opera* is reprinted (p. 976) the letter published in the 1705 *Histoire de l'Académie des Sciences*: 'Véritable Hypothèse de la Résistance des Solides, Avec la Démonstration de la Courbure des Corps qui font ressort.' This letter of 14 pages contains the solution of the problem of the elastic bending of a cantilever, in which Bernoulli obtains the correct shape but the wrong numerical constants. His diagram is very similar to Coulomb's fig. 6; Bernoulli continued to assume (with Galileo) that rotation occurred at the root about the lowest point of the beam, so that his solution did not satisfy horizontal equilibrium. Coulomb was interested in the *strength* of a cantilever, and Bernoulli in its elastic properties.

PHILIPPE DE LA HIRE (1640–1719). Proposition 125, p. 465, of the *Traité de Mécanique* (1695, later republished in 1730 in vol. 9 of the *Mémoires de l'Académie Royale des Sciences*), deals with the weight that should be given to each voussoir of an arch so that all voussoirs are in equilibrium, even though they are smooth and can slide freely one on another. 'C'est une des plus difficiles questions qu'il y ait dans l'Architecture, que de sçavoir la force qu'on doit donner aux murs & aux piédroits qui soutiennent des voutes & des arcs, pour résister à l'effort que font les voussoirs qui les forment, pour les écarter.' Thus the problem La Hire wishes to solve is that of the design of the piers or abutments of an arch, so that they stand firm under the thrust of the voussoirs, that is, the value of the arch thrust is required. La Hire's conceptual model of the arch, in which the voussoirs can slide without friction, leads him to investigate the fruitless inverse problem of the conditions for stability of the *model*.

He solves this problem by constructing a force polygon involving the weights of the voussoirs and the corresponding funicular polygon for the arch. For an arch of given shape with smooth voussoirs (La Hire takes the semi-circular arch as an example), the funicular polygon can be drawn immediately; the force polygon can then be constructed and hence the weights of the voussoirs determined. Finally, the position of the pole of the force polygon determines the horizontal thrust of the arch. Since the springing lines of a semi-circular arch are horizontal, La Hire deduces that the weights of the voussoirs at the springings must be infinite, and hence concludes that such an arch composed of smooth voussoirs cannot stand. However, he notes that in practice voussoirs cannot slide on one another, so that it is not necessary to follow exactly the theoretical calculations, but only to use them as a guide.

La Hire returned to the problem 17 years later, and used almost the same preamble in his 1712 *Mémoire* on arches: 'C'est un probleme des plus difficiles qu'il y ait dans l'Architecture, que de connoître la force que doivent avoir les pieds-droits des Voûtes pour en soûtenir la poussée...'; but here, apparently dissatisfied with the treatment based on smooth voussoirs, he proposes a theory more closely representative of the actual arch. La Hire remarks that when

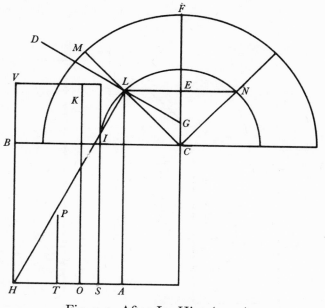

Fig. 2.3. After La Hire (1712).

the piers of an arch are too weak to carry the thrust, the arch breaks at a section somewhere between the springing and the keystone. In his fig. 1 (fig. 2.3 here), he takes the joint *LM* to be critical, and the block *LMF* is then regarded as a single voussoir, as is the block *LMI* resting on the pier *IBHS*. Thus, in fig. 2.4, by considering vertical equilibrium of the top block, the thrust *P* can be found, and La Hire assumes that this thrust acts tangentially to the intrados at the point *L*. Then, by taking moments about *H* for the lower portion of the arch and the pier, an equation expressing the stability of the whole structure may be derived.

The construction lines in La Hire's figure give certain similar triangles from which the stability equation may be deduced semi-graphically. The fact that the thrust line is tangential to the intrados at the critical section seems a 'natural' assumption, but La Hire does not comment on this. Nor does he give any rule for locating the critical point *L*, i.e. for determining the angle *LCE*.

The same ideas are then applied to the case of the plate-bande. As Coulomb points out (concluding remarks, p. 40; p. 69), it may be possible to find a more critical case for some flat arches. The state-

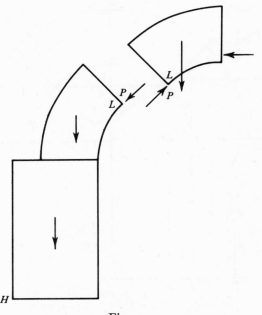

Fig. 2.4

ment that the upper 'voussoir' *LMF* remains solid is not subjected to any scrutiny by La Hire; it may in fact be impossible for some flat arches (e.g. for a thin plate-bande) to continue the line of thrust in such a way that it stays within the masonry. That is, Coulomb's second equilibrium condition has not been shown to be satisfied for the central portion of the arch between *L* and the keystone.

BERNARD FOREST DE BÉLIDOR (1697–1761) wrote several engineering texts, among them the octavo *Dictionnaire portatif de l'ingénieur* (1755), and the *Architecture Hydraulique* (1737–53) in four volumes.

The 1729 *Science des ingénieurs* referred to by Coulomb is also in the nature of an engineer's handbook. Although published in quarto, the six books of between about 60 and 100 pages each are numbered separately, and have individual decorated first pages.

Book 1 of 80 pages has the running title *de la Théorie de la Maçon-nerie*. The first three chapters deal with elementary notions of statics, and it is shown how the thickness of a retaining wall can be determined to resist a given overturning force due to earth pressure. The actual values of soil thrusts are found in chapter 4: 'De la maniere de calculer la poussée des Terres que soûtiennent les revête-mens des Terrasses & des Remparts, afin de savoir l'épaisseur qu'il faut leur donner.' Bélidor starts by noting that newly-turned (*nouvellement remuées*) and uncompacted soils (*sans être battuës*) have a natural slope of about 45° on average (less steep for sand, and steeper for clayey soils (*grasses et fortes*)). From this experimental observation he deduces that the function of the retaining wall is to sustain the pressure of a 45° prism, that is, it prevents a wedge of soil slipping along a plane of this slope. He computes the fluid pressure of a 45° triangular wedge, and then states that the *tenacité* of the soil will reduce the thrust to half the calculated value. Thus his final value for the soil thrust is one-half of the weight of the triangular wedge ($\frac{1}{4}\gamma h^2$, where γ is the soil density and h the height of the retaining wall), acting at a distance $\frac{1}{3}h$ above the wall footing. Assuming that the density of masonry is 50 per cent greater than that of the soil, it is then an easy matter to find the necessary dimensions of a retaining wall of any required shape.

Bélidor notes Vauban's *profil général* for all walls from 10 ft to 80 ft in height. These 'universal' dimensions were designed to allow for a great variety of practical conditions; the ridge thickness was specified at $4\frac{1}{2}$ ft in good quality stone, or $5\frac{1}{2}$ ft (or more) for poorer stone, with a standard batter of 1 in 5 (cf. Coulomb's remarks on p. 20; p. 54). Bélidor states that most engineers regard the batter as too great, but he himself gives tables using this value for wall heights of from 10 to 100 ft in steps of 5 ft. His ridge width for a wall of height 35 ft is 5 ft 6 in 11 lignes (cf. Coulomb's value of 5 ft on p. 20; p. 54). Further, his widths for walls of height 10, 15 and 20 ft are 1 ft 9 in 1 *l*, 2–6–2, and 3–3–5 respectively, and these values support Coulomb's remarks on p. 20; p. 55.

Bélidor gives some attention to the factor of safety, and, in a numerical example, he increases the dimensions of the retaining wall to give a safety factor of 1.2 (cf. Coulomb's value of 1.25 on p. 20; p. 54). Calculations are also made for walls with parapets and for walls with buttresses (Vauban's tables were intended *only* for buttressed walls); standard spacings for the buttresses are taken at 15 ft and 18 ft.

Book 2 of 64 pages has the running title *de la Mécanique des Voûtes*. The first chapter deals with voussoir arches and their thrusts, and starts by rejecting the empirical rule of Blondel. This rule is illustrated in fig. 2.5. The intrados of the arch is divided into three equal chords *AC*, *CD* and *DB*; *DBE* is a straight line with *DB* = *BE*, and the point *E* locates the outer edge of the supporting pier. Bélidor remarks that the rule does not involve the thickness of the arch, nor the height of the piers. (The second criticism, at least, is not perhaps of importance; Moseley demonstrated in 1843 that a finite width could be assigned to the piers to carry a given thrust, independently of their height.)

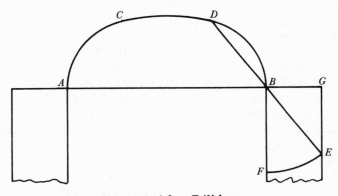

Fig. 2.5. After Bélidor.

However, Bélidor is trying to study the *mechanics* of arches, and is not content with empirical rules unless these are the result of a mathematical investigation. In fact, his work on arches is based firmly on the treatment of La Hire (fig. 2.3). Bélidor assumes that the weakest section of the arch is at 45° from the crown, and simple statics gives the value of the thrust at section *LM* as $\sqrt{2}W$, where *W* is the weight of the 'voussoir' *LMF*. However, he does not follow La Hire exactly. Besides introducing the value of 45°, he does not take the thrust to act through the point *L* on the intrados, but at the midpoint of *LM*; this leads to slightly thicker piers for the same value ($\sqrt{2}W$) of the thrust. As an exception, the thrust *is* taken to act through *L* for an example of an elliptical arch; the point *L* itself is taken to be half way round the ellipse from the springing to the keystone.

Chapters 2, 3 and 4 then deal with the calculation of the dimensions necessary for the supporting piers for various shapes of arch; a passing reference is made to the catenary and to the inverted catenary. In one calculation for which the width of the pier is found to be 6 ft 6 in 7 *l*, Bélidor remarks that this is the dimension just adequate for stability; a safety margin would be provided either by adding buttresses or by increasing the width of the pier by 5 or 6 in to about 7 ft. (This increase not only makes the pier heavier; the lever arm is also increased, so that a 7 ft pier will have an overturning moment 16 per cent greater than a 6 ft 6 in pier.)

Book 3 of 96 pages, *de la Construction des Travaux*, is a qualitative discussion of the properties of building materials, and detailed notes are given on stone, brick lime, sand, pozzolana, plaster, and mortar. Unit weights are given for a range of materials, including metals (iron, brass, copper, lead), various sands and clays, brick, various building stones, and several kinds of wood, but no other quantitative properties are given. This section also contains a much fuller set of tables than those given in book 1 for the dimensions of retaining walls with buttresses, with batters of from 1 in 5 to 1 in 10 (in unit steps) with heights from 10 ft to 100 ft in steps of 5 ft. The tables are an addendum to book 1, and were intended by Bélidor to relieve the engineer from the tedium of making his own calculations.

Book 4 consists of 104 pages on different types of civil and military buildings (*de la Construction des Edifices Militaires & civils*), but starts by presenting some experimental evidence on the strength of wooden beams. Eight experiments, each on three identical specimens, were made by 20 or 25 Artillery Officers of *l'École de la Fere*, '. . . who presented themselves at the Arsenal of that place, to convince themselves of what they had heard me say about the strength of timber . . .'. The tests were made on both simply-supported beams and on beams with fixed ends. Bélidor notes that the results of the tests agree *assez bien* with those of Parent, and he refers specifically to Parent's *Mémoires* presented to the *Académie*. By implication, Bélidor did not know of Parent's collected *Essais et recherches . . .* of 1713, in which the first completely correct solution of the bending problem is given.

Bélidor himself continues to place the neutral axis in the surface of the beam, and confirms Galileo's finding that the strength of a simply-supported beam is proportional to the square of the depth. In theory also, Bélidor states that fixed-ended beams should fail by fracture at both ends and at the centre; in this case, he believes that the load should 'exercise a third of its weight at each section which is liable

7   *87*

to fracture', which would apparently make the beam three times as strong as a simply-supported beam. (However, he does not state explicitly this factor of 3.) In the actual tests Bélidor did not manage to break the beams at their ends, and he remarks that slightly imperfect fixing would account for this. He is convinced, however, of the virtue of providing fixed ends wherever possible; from his own tests, he concludes that it would be safe to allow an increase of 50 per cent in the value of the (point) load over the simply-supported case.

Book 5 of 80 pages, *de la Décoration*, is an 'architectural' treatment of buildings, dealing with such subjects as the classical Orders, taper and entasis of columns, and so on. Book 6, also of 80 pages, *de la maniere de faire les Devis*, deals with the preparation of specifications and contracts.

3

The Strength and Stiffness of Beams

Technical biographers of Coulomb (e.g. Hollister, Hamilton) usually claim for him that he was the first to place correctly the neutral axis of bending at the centre of the (symmetrical) cross-section. Similarly Todhunter and Pearson offer only a diffuse pamphlet of 1748 by Belgrado as in any way anticipating Coulomb's discovery. However, as Saint-Venant notes in his introduction to Navier's *Résumé des leçons...*(and as remarked also by Timoshenko and by Straub), Parent had already in 1713 given a correct solution to the problem of the position of the neutral axis; Saint-Venant believes that Coulomb probably did not know of Parent's work.

Confusion arises from attributing the wrong motivation to Coulomb's work on bending. With the hindsight of the 'correct' elastic solution, in which the neutral axis for a symmetrical section is indeed centrally placed, it would seem that Coulomb was trying to solve precisely this elastic problem. However, he states quite clearly in the heading to article 7 (pp. 8–10; pp. 46–7), that he is discussing the *fracture* of bodies. He presents two theories, one for wood and one for stone. In the first, following Bossut, he assumes that at fracture the stresses are proportional to distance from the neutral axis, and, almost by accident, he obtains what is now thought of as the solution to the elastic problem. For stone he makes the alternative assumption that all the 'fibres' fracture in tension at the same stress, and he immediately moves the neutral axis to the bottom of the beam, the position which, as will be seen below, it had occupied in Galileo's time. (Coulomb notes on p. 10; p. 47 that a mathematical line would be subject to infinite compressive stresses, so that the point of rotation cannot be exactly at the bottom of the beam.)

Coulomb's two problems, of 'elastic' fracture, and of what might be called 'plastic' fracture of a material weak in tension, illustrate two different approaches to the question of bending of beams; the two problems were not always clearly distinguished in the seventeenth and eighteenth centuries.

The strength of beams (*1*)

GALILEO was concerned with the *strength* of beams. An early statement in the Dialogues *Discorsi e dimostrazioni matematiche intorno a due
nuove scienze*, published in 1638, makes clear his viewpoint:

Therefore, Sagredo, you would do well to change the opinion which you, and
perhaps also many other students of mechanics have entertained concerning the
ability of machines and structures to resist external disturbances, thinking that
when they are built of the same material and maintain the same ratio between
parts, they are able equally, or rather proportionally, to resist or yield to such
external disturbances and blows.

Instead, Galileo demonstrates the operation of the square–cube law,
and he gives examples of natural and artificial structures subjected
to their own weights.

Considering both cantilever and simply supported beams of rectangular cross-section, Galileo deduces that their fracture strengths
are proportional to the breadth and to the square of the depth of the
section. He supposes that, at fracture, the fibres of the beam (which
are assumed inextensible, cf. Coulomb's *fibres roides*, p. 9; p. 47) all
have the same 'resistance', so that the stress is uniform over the
cross-section. The total force is thus proportional to the area $b \times d$,
and assuming rotation to take place about the lowest point of the
beam, so that that point can be considered as a fulcrum for taking
moments, the moment of resistance is proportional to $\frac{1}{2}bd^2$. This is
the same quantity as that derived by Coulomb (in the same way)
for the fracture of stone.

As an expression for the section modulus of a rectangular beam,
whether the behaviour is 'elastic' or 'plastic', the quantity bd^2 is, of
course, dimensionally correct, and was confirmed repeatedly by
experiments made by later workers. The only question concerns the
factor of $\frac{1}{2}$, and this question does not in fact arise unless an attempt
is made to relate the resistance in pure tension to the bending resistance. (Such a relation was indeed sought, as will be seen.) If only
bending is considered, then the fracture strength of any size of beam
can be deduced from a single test on a beam of known size made of
the same material.

Thus Galileo was able to deduce the correct shape for a beam of
equal resistance, i.e. one which would fracture in bending simultaneously at all its cross-sections; for a beam of constant width,
subjected to a point load, the depth should vary according to a parabolic law. Todhunter and Pearson note that the 'problem of solids
of equal resistance led to a memorable controversy in the scientific

world', and they give references to this discussion; Saint-Venant summarizes a letter from François Blondel in 1661, and mentions the 1669 book of Marchetti. However, this digression is not of present interest.

Oravas and McLean mention some work of HUYGENS, which was unpublished at the time but which can now be found in the *Oeuvres Complètes*. In notes made between 1669 and 1689 Huygens applies Galileo's theory to an inclined beam (vol. 19, p. 69, *Rupture de poutres etc.*). Of great interest is a short calculation bearing the earlier date of 1662 (vol. 16, p. 381), in which Huygens equates the work done by external and internal forces at fracture of a simply supported beam. The internal work is the moment of resistance multiplied by the angle change at the point of fracture, which is set equal to the work done by the weight of the beam moving through the distance implied by this collapse mechanism. This 'virtual work' equation is, of course, precisely that used today for the plastic analysis of a simple beam system.

MARIOTTE seems to have been the first writer to record detailed experiments on beams; his *Traité du mouvement des eaux* of 1686 discusses, in part 5 section 2, the strength of water pipes. (As well as bending strength, he was concerned also with bursting pressures, and he deduced the correct expression for wall thickness of a pipe of given diameter to support a given pressure.) Mariotte made both tension and bending tests, and could not relate the results of the two by Galileo's formula; he concluded that Galileo's assumption of 'inextensibility' (i.e. of uniform fracture stress) was incorrect. Instead he stated that even the hardest materials (he made tests on glass and marble as well as on wood) must show some extension under load, and he assumed that the behaviour was linear–elastic and identical in tension and compression ('...il est très-vrai-semblable que ces pressemens résistent autant que les extensions...').

Further, Mariotte suggested that there was a maximum extension that any given material could tolerate, and that fracture would occur if that limit were passed. Using these ideas, and accepting for the moment Galileo's position of the neutral axis at the base of the section, Mariotte showed that the stresses should be distributed to give a triangular stress block (fig. 3.1), instead of the rectangular block assumed by Galileo. Thus the apparent section modulus is $\frac{1}{3}bd^2$ instead of Galileo's $\frac{1}{2}bd^2$.

However, Mariotte immediately abandons this analysis and observes that it can be imagined that, while the upper fibres of the beam are extended, those on the lower face of the beam are compressed. He places the neutral axis in the correct position, and thus, although Parent was the first to give a proof, Mariotte must be credited as the first to obtain the correct solution to the problem of elastic bending (remembering always that Mariotte was actually attempting to solve the problem of *fracture*). In his calculation of the bending resistance, Mariotte, by a singular inadvertance (as Saint-Venant remarks), dropped a factor of 2, and obtained his previous

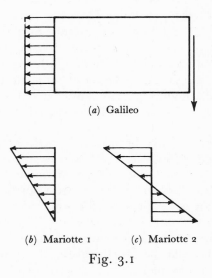

(a) Galileo

(b) Mariotte 1 (c) Mariotte 2

Fig. 3.1

result corresponding to an apparent section modulus of $\frac{1}{3}bd^2$ instead of the correct value $\frac{1}{6}bd^2$.

In fact, the results were presented not in terms of section modulus but in terms of relative and absolute strength. The absolute strength, S say, is the maximum tensile force that can be applied to the cross-section; the relative strength in bending is obtained by multiplying S by the appropriate fraction of the depth of the section. Thus Galileo derived the relative strength as $\frac{1}{2}Sd$, and Mariotte as $\frac{1}{3}Sd$; on the same basis, the correct value should be $\frac{1}{6}Sd$.

Mariotte made specific tests to check this theory, using for the purpose cylindrical specimens of $\frac{1}{4}$ in diameter made from dry wood. From a tension test he determined the absolute strength as 330 lb. He then tested the circular rods as cantilevers of length 4 in, applying (without comment) the theory for the rectangular beam. (Using the same theory for the circular rather than rectangular cross-section, the relative strength should be $\frac{1}{8}Sd$.) Thus table 3.1 may be constructed. Had he not omitted the factor of 2, Mariotte's theoretical breaking load would have been 3.4 lb. As it was, his incorrect value of 6.9 lb was reasonably close to the observed value of 6 lb, and he attempted to explain the discrepancy by assuming that the absolute strength of 330 lb had been overestimated; a load of 300 lb might have broken the specimen if applied for a long enough time.

Table 3.1

	Relative strength (in lb)	Breaking load (lb)
Galileo ($\frac{1}{2}Sd$)	41.25	10.3
Mariotte ($\frac{1}{3}Sd$)	27.5	6.9
'Correct' ($\frac{1}{8}Sd$)	10.31	2.6
Observed	(24)	6

Mariotte had, as a result of his experiments, a convincing refutation of Galileo's theory, but, due to his error in calculation, only fortuitous support for his own theory. If the fracture strength in bending of a beam can be predicted from the elastic distribution of strain coupled with a maximum-strain postulate of fracture, then it will be seen from table 3.1 that the 'correct' load that Mariotte should have observed is 2.6 lb.

Mariotte also tested a circular glass rod as a simply supported beam over a 9 in span; the central breaking load was 1 lb 10 oz 5 drams. He took a similar rod, carefully bound the ends with twine before inserting them in mortises, and repeated the test; the breaking load of this 'fixed-ended' beam was 3 lb 5 oz 4 drams, almost exactly double the simply supported value. He concluded that an encastré beam has twice the strength of the corresponding simply supported beam; this result is, of course, in exact agreement with the conclusion of modern plastic theory, even though the material, glass, has not the ductility of mild steel for which plastic theory was developed.

This work was not published until 1686, after Mariotte's death, but it was known by 1680, and in July 1684 LEIBNITZ published a paper on the beam problem in the *Acta Eruditorum* of Leipzig. Leibnitz notes that Mariotte's experimental results give breaking loads much smaller than those predicted by Galileo, but he contributes little that was new. He agrees with Mariotte that elastic strains must be considered, and he also places the neutral axis at the bottom of the cross-section. It seems clear that Leibnitz had very little real interest in the problem, essentially because the mathematical content was small; indeed, his closing remarks indicate his faith in the power of mathematics to solve physical problems: 'ut proinde his paucis consideratis tota haec materia redacta sit ad puram Geometriam, quod in physicis et mechanicis unice desideratur'.

Similarly proposition 126 ('De la résistance des solides') of LA HIRE's *Traité de Mécanique* of 1695 adds nothing to Mariotte's findings, and the neutral axis is still placed at the bottom of the section. Galileo's work on solids of equal resistance is reproduced, and the only new discussion concerns the bending of a rectangular cross-section when the neutral axis is not parallel to the faces of the beam (e.g. a square section bent about an axis parallel to its diagonal). However, this problem is dealt with only in general terms.

JAMES BERNOULLI's bending solution has already been referred to (chapter 2) and will be discussed more fully later in the present chapter. In the 1705 *Mémoire* ('Véritable Hypothèse...') he claims to be the first to consider compressions as well as extensions, whereas Mariotte had in fact discussed this matter twenty years earlier. He repeats Mariotte's precise mistake with the factor of 2, which leads him to conclude that, in general, the neutral axis can be placed anywhere, that is, the position is indifferent. James Bernoulli's *Mémoire* of 1705 is a revised version of his 1694 Leipzig paper; the later work was written in the light of a *Mémoire* by VARIGNON in 1702.

Varignon continues to place the neutral axis at the base of the section, but develops what might be termed a unified theory of bending. He obtains a single formula for the moment of resistance of a beam in bending in which the fibre stress is given as a general function of the fibre strain. Thus, if the fibre stress is taken to be constant (i.e. independent of the strain), he obtains Galileo's solution. As an interesting general example, he discusses the case $\sigma = k\epsilon^m$ as representative of a non-linear stress–strain law, but, beyond making the integration, he carries the work no farther.

The work of PARENT was almost completely ignored at the time, and continued to be little known. He published a series of *Mémoires* in the *Académie* in the first decade of the eighteenth century, but never advanced beyond the lowest grade of *élève* (see chapter 7 for some notes on the organization of the *Académie*). He collected and expanded his papers in three volumes of *Essais et recherches...* in 1713. The official *Éloge* (*Histoire de l'Académie...*, pp. 88–93, 1716) states that these volumes are 'plein de bonnes choses', but that they had no great success, and wryly advances several reasons for this, among them the fact that the books are inconvenient, being both very small (duodecimo) and very thick.

Parent's earlier papers on bending continue to place the neutral axis at the base of the section, but he reviewed the whole theory in the collected essays. The *Mémoire* in volume 2, pp. 567ff, is entitled 'Comparaison des résistances des Cylindres & segmens pleins, avec celles des creux égaux en base, dans le systême de M. Mariotte'. Parent comments first on Mariotte's use of theory for a rectangular section to explain experimental results on a beam of circular cross-section. By making the integration over the circle (still with the neutral axis at the base), Parent obtains the value for the relative strength, $\frac{5}{16}Sd$, instead of the value $\frac{1}{3}Sd$ taken by Mariotte, and he justly observes that had Mariotte used the slightly smaller value, the theoretical result would have been closer to that observed, table 3.2. (Parent in fact states that Mariotte would have found the actual observed value.)

Table 3.2

	Relative strength (in lb)	Breaking load (lb)
Observed	(24)	6
Parent (1): $(\frac{5}{16}Sd)$	25.78	6.4
Parent (2): $\frac{9}{11}(\frac{1}{3}Sd)$	22.5	5.6

However, Parent had already discovered and corrected Mariotte's mistake of the factor 2 in the value of the relative strength calculated with respect to a central neutral axis; thus he had already arrived at the value $\frac{1}{6}Sd$ for a rectangular section, and notes that Mariotte should have calculated the (wrong) value of $3\frac{1}{2}$ lb as the breaking load, 'ce qui seroit bien éloigné de la verité'.

Later in the same *Mémoire*, Parent discusses the position of the neutral axis at the instant of fracture. Just before fracture the extreme tensile fibre will be the most highly strained, and the behaviour of that one fibre will govern the fracture of the whole section. It is true that the neutral axis will descend during fracture until it reaches the base of the section, but it will never be found there *before* fracture.

Thus he distinguishes clearly between the elastic working state of the beam, and the ultimate condition which is governed by the weak tensile behaviour of the material, and he is aware that the neutral axis can shift between one state and the other.

Parent finally and explicitly locates the neutral axis in the 14th *Mémoire* of volume 3 of the *Essais*: 'De la véritable méchanique des résistances rélatives des Solides, & réfléxions sur le Système de M. Bernoulli de Bâle.' (This volume contains 32 *Mémoires*, of which 'about one third' had been read to the *Académie*.) Parent states that the total *résistance* of the fibres of the compressive triangle *CBX*, fig. 3.2, must equal that of the tensile triangle *ACT*, these two forces acting through *I* and *D* respectively, 'qui est une proprieté dont personne n'avoit encore parlé'. Thus he is completely aware that his correct statement of horizontal equilibrium is breaking new ground.

His general strain distribution permits unequal elastic moduli in tension and compression; the tensile stress in fibre *AT* will govern fracture, but the compressive stress in the fibre *BX* may well be much larger. He shows that the relative strength in bending for a rectangular section is directly proportional to *AC*, i.e. to the distance of the neutral axis from the extreme fibre of the beam. He applies this result to Mariotte's test, apparently making the same mistake as Mariotte in using a result deduced for a rectangular section to explain an experiment on a circular rod; he deduces that $AC/AB = 9/11$ in Mariotte's test, that is, that the neutral axis lies very close to the bottom fibre. A relative strength of $\frac{9}{11}(\frac{1}{3}Sd)$ does not in fact agree exactly with the experimental result, table 3.2, but the passage is slightly obscure and the figure of $\frac{9}{11}$ is in any case suspect; it is given in the text as $\frac{11}{9}$, which is impossible, and corrected to $\frac{9}{11}$ in the *Errata*.

Parent corrects Bernoulli (fairly politely for Parent; his apparent rudeness contributed to his lack of academic success) in saying that Mariotte had not considered compression on the bottom face of the beam. BÜLFFINGER makes the same remark, and quotes the earlier *Mémoires* of Parent, but does not seem to be aware of the latter's *Essais*. Bülffinger's *Mémoire* of 1729 contains little new theory, but his survey of previous work is reasonably complete, and his conclusions, if unsatisfactory in that no definite result emerges, seem unexceptionable. Mariotte's theory (neutral axis at the centre) is not accepted by Bülffinger, since experimental evidence is lacking; similarly, there is no evidence that the extension and compression of the fibres is linear–elastic. He revives Varignon's idea that stress might be proportional to some power of the strain; however, until this stress–

strain law has been determined experimentally, theory cannot fix the position of the neutral axis.

Thus both Parent and Bülffinger, in critical and closely-reasoned essays, were moving towards the conclusion that the calculation of the bending strength of a beam was not necessarily a simple matter. Parent considered linear–elastic behaviour with unequal tensile and compressive moduli, and Bülffinger a non-linear stress–strain relation. Certainly, Mariotte's experimental results were becoming less, and not more, explicable in terms of a simple elastic theory of bending.

Neither Bélidor nor Musschenbroek, both of whom wrote in the same year (1729) as Bülffinger, and whose work has already been noted in chapter 2, contributed to the theory of bending, but their experimental results added very greatly to the data available for analysis. In the next 40 years (up to the time of Coulomb's *Essai*) many tests on wood, stone and iron are reported. Among these may be noted those of BUFFON in 1740 and 1741; he gives results of a very large number of full-scale tests on wooden beams. POLENI (1748) tested the strength of iron in his investigation of the reinforcement required for the dome of St Peter's (see chapter 6). In this same period of 40 years, however, work was more concerned, as will be seen later in this chapter, with the *flexure* of beams rather than with their strength.

Coulomb's 'Essai'

It is against the background of the previous section that Coulomb's contribution to the theory of beams must be assessed. His experiments were few and not as carefully made as those, for example, of Musschenbroek; only the direct shear test seems new, and this experiment is somewhat imprecise (see chapter 4). Coulomb's experimental work did not, in fact, add much to existing knowledge.

The significant passages in Coulomb's article 7 are the very clear statements of the equilibrium conditions that must be satisfied at the cross-section of the beam (p. 8; p. 46). Thus not only must the tensile forces balance the compressive forces; vertical equilibrium must also be achieved, that is, the cross-section of the beam must carry a shearing force. Further, the resisting moment of the cross-section must equal the bending moment of the external loading. Each of these three equilibrium statements had in fact been recognized and made by Parent sixty years earlier, and Bülffinger also was aware of the conditions, but Coulomb was very much clearer in his presentation.

Of the three master statements of structural mechanics, those of equilibrium, compatibility and of the stress–strain law of the material,

Coulomb really deals only with the first; indeed, he states in his opening sentence that he is dealing with problems of statics. In his discussion of elastic bending (p. 9; pp. 46ff), he does in fact assume implicitly that plane sections remain plane, and he proceeds to determine the stress distribution by the explicit use of a linear stress–strain relationship. However, any faint idea of compatibility is abandoned immediately afterwards in the derivation of Galileo's formula. There is, of course, an infinite number of stress distributions which will satisfy Coulomb's three conditions of equilibrium, and much more sophisticated attempts to solve the problem had already been made (almost certainly unknown to Coulomb) by Parent and by Bülffinger.

All that Coulomb offers in his remarks on fracture is that Galileo's formula ($\frac{1}{2}Sd$) should be used for stone, and Parent's ($\frac{1}{6}Sd$) for wood. Nothing new is proposed, and Coulomb offers no hint that he might have grasped that the elastic formula could be used for stone *before fracture*.

It is difficult to escape the conclusion that, as an engineer, Coulomb was not really interested in the ultimate *strength* of masonry construction. He deals with the breaking of beams and with the fracture of columns, but the first engineering *design* in the *Essai*, that of a retaining wall, pp. 19ff; p. 54, does not involve the strength of the material at all. Instead, the wall is designed to resist the lateral thrust purely by the moment of its dead weight. Similarly although Coulomb introduces cohesion into his analysis of arches, p. 35; p. 66, it is almost immediately neglected (p. 38; p. 68) and the stability of arches is shown to be a matter of geometry rather than of strength of materials (see chapter 6). (Coulomb also ignores cohesion in his two specific calculations of soil thrust, pp. 19 and 22; pp. 54 and 56.)

Twenty-five years after Coulomb's *Essai*, in 1798, GIRARD's book, while referring to Coulomb, still continues to place the neutral axis at the bottom of the cross-section; he could find no use for an elastic theory, based upon the true position of the neutral axis, as an explanation of the experimental results (Girard reports many further tests on wooden beams). It is difficult here to separate ignorance from expedience. When Galileo and Mariotte placed the neutral axis at the base of the section they were almost certainly ignoring the condition of longitudinal equilibrium. When Coulomb also placed the neutral axis at the base he was aware (as Parent would have been aware) that this implied infinite compressive stresses, if longitudinal equilibrium was to be maintained. But when Girard made the same assumption about the position of the neutral axis it may have been purely to give some semblance of theory to the empirical constants of Galileo and Mariotte that fitted best the experimental results. In any case it was clear, whether or not a particular writer was aware of the necessity of satisfying equilibrium conditions, that an elastic modulus of $\frac{1}{6}bd^2$, derived from a central neutral axis, was useless as an explanation of fracture; the value of $\frac{1}{2}bd^2$ (Galileo) seemed best for stone, and $\frac{1}{3}bd^2$ (Mariotte) for wood.

Girard gives a 47-page historical introduction to his subject, and mentions the names of Galileo, Euler, Lagrange, François Blondel,

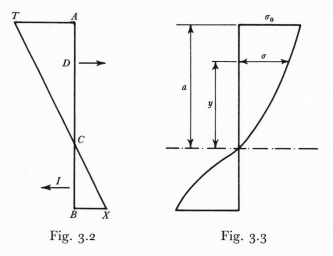

Fig. 3.2   Fig. 3.3

Marchetti, Mariotte, Leibnitz, Varignon, James Bernoulli, Coulomb, Parent, Bélidor, Mussenbrock [*sic*], Buffon, and others, discussing in each case their contributions to the problem. In the case of Parent, reference is made to the *Mémoires* of 1707 and 1708 of the *Académie*, and not to the collected *Essais* of 1713. Thus it seems fairly certain that Girard did not know of Parent's final work on bending, and, by implication, equally certain that Coulomb had arrived at an independent solution.

It is of the utmost interest that Coulomb read Girard's book before it was published. There is a short foreword by Prony (who was one of the founders of the *École Polytechnique* in 1794, and who became Director of the *École des Ponts et Chaussées* in 1798). Prony states that he and Coulomb were asked to report on Girard's work, and he gives a summary of the contents. In approving the book, Prony states that they regard it as the most complete work on strength of materials both from the point of view of experiment and of theory.

Coulomb adds his signature to this statement. Thus Coulomb himself supported the view that calculations for the fracture of beams should be based on the neutral axis lying at the bottom of the section. He was unaware that he had in fact solved the problem of *elastic* bending of beams, in which the neutral axis is central.

The strength of beams (2)

BARLOW, in the first edition of his book on the strength of timber (1817), did not know of Coulomb's work. He rejects the theories of Galileo and of Mariotte on experimental grounds; his tests of wooden beams (typically 2 in square in section and 48 in long) showed that the neutral axis did not lie in the surface. Instead, the beams showed fractures in which about $\frac{3}{8}$ of the section had failed in tension, and the remaining $\frac{5}{8}$ in compression. He bases some spurious analysis on this observation, but corrects his mistaken theory in the 1837 edition of the book, and adds:

Fortunately it is seldom that the strength of timber is of great importance, except in the form of rectangular or square beams; and its strength in these forms is deducible from experiments on similar formed beams, without any reference to the exact position of the neutral axis; but still, as a point of theory, and wherever the question relates to beams of other figures, it is essential that we should have reference to it.

Similarly TREDGOLD, in 1822, had curious theoretical ideas about the position of the neutral axis, and it fell to HODGKINSON, in two Memoirs published in 1824 and 1831, to make the next theoretical advance in the problem of fracture in bending. Hodgkinson knew the French work, and, in particular, he refers to Coulomb, and he constructed a general stress distribution to satisfy the equilibrium conditions. In effect, Hodgkinson stated that the tensile stress might be represented by the formula

$$\sigma = \sigma_0 \left(\frac{y}{a}\right)^n,$$

where σ_0 is the extreme fibre stress, and a is the distance of the neutral axis from the extreme fibre (fig. 3.3). A similar expression, but with different constants (a' etc.), holds for compression. For a rectangular section of depth $d = a + a'$, Hodgkinson shows that horizontal equilibrium is satisfied if

$$\sigma_0 a / (n+1) = \sigma_0' a' / (n'+1),$$

and he obtains a general expression for the moment of resistance.

The constants are to be obtained experimentally and, once they are known, bending strengths can be predicted. For example, Hodgkinson finds that for Quebec oak, $n = 0.97$ (tension) and $n' = 0.895$ (compression). A bending test to fracture had given $a/a' = \frac{23}{25}$, and the value of σ_0 was taken as 8000 lb/in². Thus the moment of resistance of a given beam is calculable. (The experiments to determine n and n' were simple tensile and compressive tests; tensile strengths of some other materials are taken from Musschenbroek.)

The non-linear stress distribution is, of course, not necessarily related in any particular way with strain; Hodgkinson's assumption is purely one of equilibrium, and compatibility and stress–strain relations have not been used. His problem was to predict the breaking strength of a beam in bending by an empirical formula which embodied correct equilibrium statements. In fact, Hodgkinson assumed that strain in bending was proportional to the distance from the neutral axis, so that the assumed power law for stress was really an assumed stress–strain relationship of the form $\sigma = k\epsilon^n$. Hodgkinson's theory was in some sense a synthesis of the work of Varignon, Parent and Bülffinger; Saint-Venant, as will be seen, made some improvements, but the problem of fracture had at last been effectively solved.

Navier and Saint-Venant

The first edition (1826) of NAVIER's *Leçons...* gives for the first time the completely 'correct' *elastic* solution to the problem of bending. (References here are to the 1864 edition, edited by SAINT-VENANT.) Article 3, paragraphs 76–80, establishes the general elastic formula $M/I = E/R$. Saint-Venant notes that Navier has tacitly assumed (1) that sections originally plane remain plane after bending, and normal to the curved fibres, and (2) that individual fibres are free to expand and contract without affecting their neighbours. Saint-Venant shows that both assumptions are in general false, but that the form of the elastic formula is nevertheless valid.

Article 4, *Rupture par flexion*, paragraph 112ff, deals with the problem of fracture. Navier states that the simplest hypothesis, and the one closest to reality, is that the greatest elastic strain, in either compression or tension, governs the fracture. Thus his previous results, in which, for example, the neutral axis is shown to pass through the centre of gravity of the cross-section, may be used to determine the fracture moment. The section modulus can be written $z = I/a$, where a is the distance of the extreme fibre from the neutral axis, so that $M_0 = \sigma_0 z$. (Saint-Venant has a long note at this point, in which he examines unsymmetrical sections, and material with different properties in tension and in compression.) Navier then derives section moduli for the rectangle ($\frac{1}{6}bd^2$) and for the circle ($\frac{1}{4}\pi r^3$), and also for a rectangle bent about an inclined axis. Saint-Venant points out that this last analysis is wrong, since Navier has not considered bending to occur about the principal axes. Navier mentions the shift observed by Barlow in the position of the neutral axis (to $\frac{5}{8}$ of the depth), and he derives Galileo's and Mariotte's moduli of $\frac{1}{2}bd^2$ and $\frac{1}{3}bd^2$.

Navier finally remarks (paragraph 151) that his theory of fracture is based on the hypothesis that behaviour is linear–elastic right up to the point of failure; if experimental results are at variance with the theory, then behaviour in tension must differ from that in compression, with a consequent shift in the position of the neutral axis. He points out, however, that (1) the section modulus of a rectangular section is always of the form kbd^2 (Saint-Venant extends this to similar shapes of cross-section of any size) and (2) in practice, it is only elastic behaviour that is of interest, and that there is never any question of considering the state at fracture. This last statement, while perhaps reflecting general opinion in the first quarter of the nineteenth century, seems to be the first clear exposition of the elastic philosophy of design.

Navier's remarks give rise to another long note by Saint-Venant, in which he refers to the non-linear hypotheses of Varignon and of Hodgkinson. Saint-Venant proposes the slightly more convenient form

$$\sigma = \sigma_0\left[1 - \left(1 - \frac{y}{a}\right)^n\right]$$

for the tensile stress as a function of distance from the neutral axis, with a similar expression with different constants (σ_0', etc.) for compression. Equating tensile and compressive forces for a rectangular cross-section of width b,

$$\frac{n}{n+1}\,a\sigma_0 = \frac{n'}{n'+1}\,a'\sigma_0',$$

and the bending moment at the cross-section may be written

$$M = \frac{ba^2\sigma_0}{2}\,\frac{n(n+3)}{(n+1)(n+2)} + \frac{ba'^2\sigma_0'}{2}\,\frac{n'(n'+3)}{(n'+1)(n'+2)}.$$

Saint-Venant discusses several cases.

Case (i): $n = n' = 1$, fig. 3.4.

Using the result $a\sigma_0 = a'\sigma'_0$, and noting that $a + a' = d$, the expression for the bending moment reduces to Parent's form

$$M = \left(\frac{a}{d}\right)\left(\tfrac{1}{3}bd^2\sigma_0\right).$$

For the neutral axis on the centre line $(a/d = \tfrac{1}{2})$, the elastic modulus of $\tfrac{1}{6}bd^2$ is recovered. For the neutral axis dropping to the bottom of the section, $M = \tfrac{1}{3}bd^2\sigma_0$, and Mariotte's formula results, in which it is assumed that the tensile stress governs fracture, the material being able to resist very large compressive stresses.

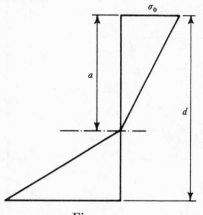

Fig. 3.4

For the other cases, Saint-Venant assumes that the slope $d\sigma/dy$ at the neutral axis is the same in both tension and compression; that is, for small values of σ, and imagining the stresses to result from small elastic strains, the elastic moduli in tension and compression are made equal. This restriction gives the further relation

$$\frac{n\sigma_0}{a} = \frac{n'\sigma'_0}{a'},$$

and the value of the bending moment may be written

$$M = \tfrac{1}{2}bd^2\sigma_0 \frac{1}{(1+k)^2}\left(\frac{n}{n'+1}\right)\left[\left(\frac{n+3}{n+2}\right) + \frac{1}{k}\left(\frac{n'+3}{n'+2}\right)\right],$$

where

$$k = \left(\frac{n+1}{n'+1}\right)^{\frac{1}{2}} = \frac{a}{a'}.$$

Case (ii): $n = n'$, fig. 3.5.

The stress distribution is symmetrical; $k = 1$ and $a = a' = \frac{1}{2}d$. Thus

$$M = \tfrac{1}{4}bd^2\sigma_0\left(\frac{n}{n+1}\right)\left(\frac{n+3}{n+2}\right).$$

The value $n = 1$ gives the linear–elastic case, i.e. $M = \frac{1}{6}bd^2\sigma_0$. As the value of n increases, so the value of M approaches $\frac{1}{4}bd^2\sigma_0$, which is the maximum moment of resistance of a beam made from perfectly plastic material. Saint-Venant sketches the stress distribution for $n = 10$ (for which the value of M is less than 2 per cent below the full plastic value) and, although he did not refer explicitly to the possibility of perfect plasticity, Saint-Venant must be credited with the first derivation of the plastic section modulus.

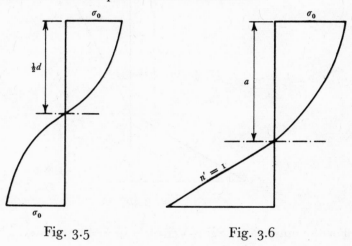

Fig. 3.5   Fig. 3.6

Case (iii): $n' = 1$, fig. 3.6.

Here the explicit assumption is that fracture is governed by the maximum tensile stress, and that the compressive distribution is of little interest. The various quantities become:

$$k = \left(\frac{n+1}{2}\right)^{\frac{1}{2}}, \quad a = \left(\frac{k}{1+k}\right)d,$$

and

$$M = \tfrac{1}{4}bd^2\sigma_0\frac{1}{(1+k)^2}n\left[\frac{n+3}{n+2}+\frac{4}{3k}\right].$$

As before, $n = 1$ gives the elastic case $M = \frac{1}{6}bd^2\sigma_0$. As n increases, so the value of a approaches d, and, in the limit, Saint-Venant notes that Galileo's formula $M = \frac{1}{2}bd^2\sigma_0$ is obtained.

Further, Saint-Venant points out that the section modulus always lies between the limits $\frac{1}{6}bd^2$ (Coulomb) and $\frac{1}{2}bd^2$ for n, $n' > 1$. He proposes, therefore, that the formula $M = \frac{1}{6}\alpha bd^2\sigma_0$, where α lies between 1 and 3, will serve as an empirical expression for predicting rupture moments. Thus for cast iron α is about 2, which gives a value of n in case (iii) above of between 5 and 6, and the theory can then be applied to the bending of non-rectangular sections, using these numerical constants.

Thus, while not allowing for a falling stress–strain characteristic, Saint-Venant derived self-consistent empirical expressions from which the strengths of beams in bending could be calculated. Todhunter and Pearson object that Saint-Venant offers no experimental evidence that values of n and n' deduced from a test on one shape of cross-section would be the same as those deduced from any other test. For the symmetrical case (ii), however, this objection does not arise, and modern plastic calculations are nothing more than the limiting condition (n large) of this case.

The stiffness of beams

It was mentioned above that for the half-century before Coulomb's *Essai* little work was being done on the strength of beams. However, it was in this period that Euler and the Bernoullis were investigating the difficult and interesting problem of the *shape* of a bent elastic member. Galileo had mentioned the problem of the deflected form, but it was JAMES BERNOULLI who made the first step in the solution of the problem. In 1691 he published a logogryph: *Qrzumu bapt dxqopddbbp...*; the secret of the simple letter substitution (a letter being replaced by the next in the alphabet, the second by the letter three away, and the third by the letter six away, e.g. *aaaaa...* would be encoded *bdgbd...*) he revealed in 1694, and interpreted his result: 'Portio axis applicatam...' His statement was that the radius of curvature at any point of an initially straight uniform beam was inversely proportional to the bending moment at that point.

This was amplified in the *Explicationes* of 1695, in which Bernoulli considers the relative inclination of two neighbouring cross-sections of the beam. The neutral axis is taken at the bottom of the section, but this does not affect the result that curvature is proportional to bending moment. Bernoulli does not make the assumption of small slopes, and hence the equation of the elastic curve is not obtainable in terms of elementary functions. However, his equation is of first order only, and hence easily soluble in terms of a series expansion. (Saint-Venant shows that, by a suitable choice of axes, the general second-order differential equation of bending can always be integrated once without approximations). Finally, in 1705, Bernoulli returned to the subject of beams ('Véritable Hypothèse...') in a letter, already noted, to the French *Académie*.

DANIEL BERNOULLI wrote to Euler in 1742, during the course of a long and lively correspondence in the lingua franca in which scholars communicated:

Ew. reflectiren ein wenig darauf ob man nicht könne sine interventu vectis die curvaturam immediate ex principiis mechanicis deduciren. Sonsten exprimire ich die vim vivam potentialem laminae elasticae naturaliter rectae et incurvatae durch $\int \mathrm{d}s/R^2$, sumendo elementum $\mathrm{d}s$ pro constante et indicando radium osculi per R. Da Niemand die methodum isoperimetricorum so weit perfectionniret als Sie, werden Sie dieses problema, quo requiritur ut $\int \mathrm{d}s/R^2$ faciat minimum, gar leicht solviren.

The 'vis viva potentialis laminae elasticae' is, to within a constant, the strain energy in bending; Daniel Bernoulli had found that this quantity was a minimum for the elastic curve of his uncle James. He proposes that Euler should apply the calculus of variations to the inverse problem of finding the shape of the curve of given length (and satisfying certain end conditions) so that the function $\int \mathrm{d}s/R^2$, where R is the radius of curvature, is a minimum.

EULER makes this analysis in the 'Additamentum I, De Curvis Elasticis' to his *Methodus inveniendi lineas curvas...* of 1744; following Daniel Bernoulli's suggestion, he quickly derives the first-order differential equation of the elastica. This derivation is purely mathematical; Euler then shows that precisely the same differential equation arises from James Bernoulli's result that bending moment is proportional to curvature. The external load can effectively be reduced to a single force P, and Euler classifies his solutions according to the angle at which this single resultant cuts the elastica (at a point of inflexion, of course). With axes x and y, fig. 3.7(a), Euler's differential equation is

$$\frac{\mathrm{d}y}{\mathrm{d}x} = \frac{(a^2 - c^2 + x^2)}{(c^2 - x^2)^{\frac{1}{2}} (2a^2 - c^2 + x^2)^{\frac{1}{2}}},$$

where $a^2 = 2EI/P$, EI being the flexural rigidity (which Euler calls the absolute elasticity, and denotes Ek^2). The constant c arises from a first integration, and the form of the solution of the equation is clearly governed by the relative magnitudes of a and c. At the origin, the slope of the elastica is

$$\left[\frac{dy}{dx}\right]_{x=0} = \frac{a^2 - c^2}{c(2a^2 - c^2)^{\frac{1}{2}}};$$

for $a = c$, for example, the slope is zero, while for $c^2 > 2a^2$, the curve cannot cut the y-axis at all.

Euler distinguishes nine classes of solution, of which his classes 2, 4, 6 and 8 are sketched in figs. 3.7(a) to (d); classes 3, 5 and 7 are unique solutions representing transition curves between those sketched in fig. 3.7, while classes 1 and 9 represent the end points of the analysis.

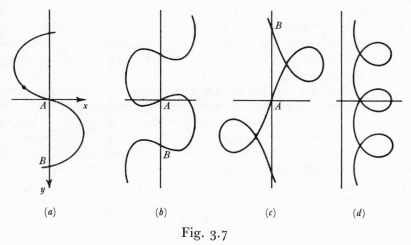

(a) (b) (c) (d)

Fig. 3.7

Thus in the first class the slope at the origin is supposed to be zero, and the deflexions in fig. 3.7(a) to be infinitesimal. Euler shows easily that these infinitesimal deflexions are sinusoidal, and that further they can only be maintained for a certain finite value of the force P, equal to (the 'Euler buckling load') $\pi^2 EI/l^2$. (Euler imagines a string AB connected to the elastica; this string must of necessity carry this constant force.)

For the general second class, fig. 3.7(a), the load necessary to maintain equilibrium is greater than $\pi^2 EI/l^2$; this class arises for $0 < c < a$. Class three occurs for $c = a$, and class four, fig. 3.7(b), for $1 < c^2/a^2 < 1.651868$. At the upper limit (class five), the points A and B just coincide, and the elastica takes on the form of a figure eight. In class six the points A and B cross over, until at $c^2 = 2a^2$ the solution switches to that sketched in fig. 3.7(*d*) (class eight). Finally, for *c* very large, class nine, the elastica is bent into a circle.

Euler assumed that the *elasticitas absoluta* was proportional to the square rather than the cube of the depth of the (rectangular) cross-section. However, he proposed in any case that its value should be determined experimentally, using his approximate result for the small deflexion δ of a transversely-loaded cantilever:

$$EI = \frac{Wl^2(2l - 3\delta)}{6\delta}.$$

(It may be noted that for δ ≪ 1, δ = $Wl^3/3EI$.)

Euler extends his analysis to members of non-uniform cross-section, to initially curved members, and to members subjected to distributed loading. He also discusses the free oscillation of beams.

The work is of great mathematical interest, and the classification of the results, fig. 3.7, without detailed solutions of the differential equation, is of extraordinary ingenuity. Further, the basic result of class one, that of the Euler buckling load, has of course been of the utmost importance in engineering design, despite the fact that even fig. 3.7(*a*) (to pick the least distorted picture of the elastica) would not seem to serve in any sense as a model for a practical problem. A consequence of the fact that Euler did not limit his discussion to the case of small deflexions is that the general differential equation of bending, involving curvatures

$$\frac{d^2y}{dx^2}\left[1 + \left(\frac{dy}{dx}\right)^2\right]^{-\frac{3}{2}},$$

is not integrable in terms of simple functions. In addition, the axes are taken to be in the direction of and at right angles to the resultant applied load, rather than being related to the initial direction of the beam. Thus, even for small deflexions, the simplification $(dy/dx)^2 \ll 1$ cannot be made, except for Euler's first class of axial compression, and, by an interchange of the axes, for class three (transverse loading).

Daniel Bernoulli had in fact seen in 1741, during a study of the vibrations of a cantilever beam, that the curvature for small deflexions could be expressed by d^2y/dx^2, and he obtained the linearized equation

$$EI \frac{d^2y}{dx^2} = Wx,$$

where the axes are taken at the tip of the cantilever. He integrates this equation to obtain the tip deflexion, δ = $Wl^3/3EI$.

Euler treated his classes one and three again in his Berlin *Mémoire* of 1757. He derives directly the equation

$$EI\frac{\mathrm{d}^2y}{\mathrm{d}x^2} = -Py$$

for the column, and discusses the pin-ended case. In all his work, Euler obtained only the lowest critical loads, and it was LAGRANGE who gave the first satisfactory account, in 1770–73, of the higher buckling modes.

Lagrange showed that the buckling load of a pin-ended column could be expressed as $m^2\pi^2EI/l^2$, where m is any integer, and his discussion is very much along the lines that are followed by any standard elementary text of today. He then writes the full equation, not making the assumption of small slopes, and obtains the solution in terms of a definite integral (the elliptic integral), which he solves in terms of an infinite series, rapidly convergent for small deflexions.

The lengthy remainder of the paper is concerned with the initial buckling of a straight pin-ended column which is not prismatic, but which is a solid of revolution. Lagrange shows that if the generating curve is a conic, then the most efficient conic is the straight line, that is, the column is conical for minimum material consumption. Further, he shows that the most efficient cone is the degenerate right circular cylinder. He then sets up the general variational equations for the most efficient profile, but he did not succeed in obtaining the correct solution. This particular problem lay dormant until 1851, when CLAUSEN obtained a solution involving entasis; recently (1960) KELLER has made a more general study.

Navier read a series of papers to the *Académie* in 1819, 1820, and later, in which he dealt with elastic bending of beams and plates, and his findings are summarized in the *Leçons*. He gave the first 'modern' engineering treatment of small deflexions of beams, with the curvature expressed as $\mathrm{d}^2y/\mathrm{d}x^2$. He considered also statically-indeterminate beam systems, and showed how to calculate redundant reactions. Further developments came very quickly; for example, CLAPEYRON's slope-deflexion equations, and BERTOT's method of solution (the equation of three moments), were both published before 1860.

Thus, by 1864, the year of Saint-Venant's edition of Navier's *Leçons*; the problem of the beam had been essentially solved; satisfactory accounts were available of both the elastic behaviour and the ultimate strength. What remained was to incorporate this knowledge into a more general theory of structures.

4

Coulomb's Equation

Coulomb reports three groups of tests on stone. Test 1 on p. 6; p. 45 is a tensile test, and test 2 an attempt to load a specimen in pure shear. Test 3 on p. 7; p. 45 investigated the fracture of a cantilever beam. All these tests were made on stone from Bordeaux (p. 6; p. 45) and would therefore seem to have been done in France, probably between the summer of 1772 when Coulomb returned from Martinique and March 1773 when he read the *Essai* at the *Académie*. The tests on Provence bricks (p. 7; p. 45) were presumably also done in France; the test on mortar reported on the same page was done in Martinique. Coulomb does not report any compression tests of his own, and the single result he quotes, for brick (p. 13; p. 50), is due to Musschenbroek. However, he develops a complete theory of compressive failure of axially-loaded stocky columns, and this will be discussed here.

Coulomb postulates (article 8, p. 10; p. 48) a plane of failure along which, in the limiting state, one portion of an axially-loaded column will slide over the other, and the problem he solves is the location of this plane. The force resisting the sliding is made up of two parts, namely the natural cohesion of the material, assumed constant, and the internal friction, assumed to be proportional to the normal pressure. In terms of stresses,

$$|\tau| = c + \sigma \tan \phi, \qquad (4.1)$$

where τ is the shearing stress acting on the plane of failure at the instant of fracture, and σ is the corresponding compressive normal stress; the material constants c and ϕ are the cohesion and the angle of friction respectively. The friction constant n used by Coulomb is given by

$$\tan \phi = 1/n. \qquad (4.2)$$

Figure 4.1 shows the forces acting on the top portion of the column as it fractures along a plane inclined at an angle θ to the horizontal (Coulomb's angle x at M, p. 11; p. 48); the cross-sectional area of the column (in fact of arbitrary shape) has been taken as a^2. For equilibrium,

$$\tau \cos \theta = \sigma \sin \theta, \qquad (4.3)$$

and
$$P = (\sigma\cos\theta + \tau\sin\theta)\, a^2\sec\theta. \qquad (4.4)$$

The stresses σ and τ can be eliminated from equations (4.1), (4.3) and (4.4) to give

$$P = \frac{ca^2}{\cos\theta(\sin\theta - \tan\phi\cos\theta)}, \qquad (4.5)$$

which is, of course, identical with Coulomb's expression on p. 12; p. 49. He finds that the minimum value of P occurs when

$$\tan\theta = [(1+n^{-2})^{\frac{1}{2}} - n^{-1}]^{-1}, \qquad (4.6)$$

and this expression can be written in several forms:

$$\tan\theta = (1+n^{-2})^{\frac{1}{2}} + n^{-1} = \frac{1}{\sec\phi - \tan\phi} = \sec\phi + \tan\phi$$
$$= \tan(\tfrac{1}{4}\pi + \tfrac{1}{2}\phi) = \cot(\tfrac{1}{4}\pi - \tfrac{1}{2}\phi). \quad (4.7)$$

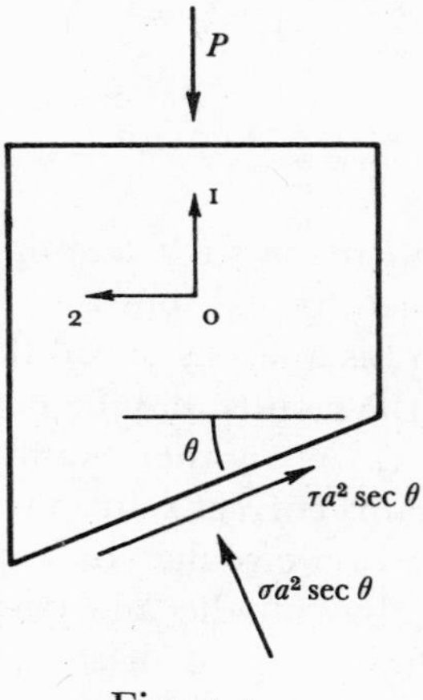

Fig. 4.1

Thus the introduction of a friction angle ϕ instead of the coefficient $1/n$ leads to a simple explicit expression for the location of the plane of fracture, namely
$$\theta = \tfrac{1}{4}\pi + \tfrac{1}{2}\phi. \qquad (4.8)$$

The substitution of this value of θ into equation (4.5) gives the value of P to cause fracture of the column:

$$P_{\min} = 2ca^2\tan(\tfrac{1}{4}\pi + \tfrac{1}{2}\phi) = 2ca^2\cot(\tfrac{1}{4}\pi - \tfrac{1}{2}\phi) = 2ca^2\cot s, \quad (4.9)$$

where $s = (\tfrac{1}{4}\pi - \tfrac{1}{2}\phi)$ is the angle introduced by PRONY.

Prony had realized as early as 1802 the advantages to be gained by working in terms of an angle of friction, and had deduced the result, 'remarquable par sa simplicité', illustrated in fig. 4.2. On a horizontal line AB is set off the friction angle ϕ to give a right-angled triangle ABC; then the bisector CD of the angle at C gives the plane of fracture. (Prony developed this result to determine the wedge of soil giving the greatest thrust on a retaining wall, see chapter 5 below; in fig. 4.2, AB is the horizontal surface of the soil and AC the vertical retaining wall.)

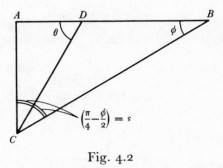

Fig. 4.2

Coulomb's failure to make such a simplification (that is, to use equation (4.2)) may seem trivial, but is to some extent typical of his work in mechanics; in his analysis of soil thrust, he uses a distance x as variable, although the results may be expressed better in terms of an angle (see chapter 5). As another example, again connected with friction, Coulomb obtains correct solutions in his *Machines Simples* of 1781 for the behaviour of a circular shaft in a slightly loose bearing; however, although he has all the analysis and even the diagrams, he fails to take the final step of discovering 'Coulomb's' friction circles. And in the present *Mémoire*, although the compressive analysis, equation (4.5), is complete and correct, there is the serious omission of any attempt to solve the corresponding tensile problem; this point is discussed below.

The greatest disappointment is, of course, the fact that Coulomb did not develop a more general theory of the equilibrium of stress. He had realized (p. 2; p. 42) that his work on the fracture of columns was essentially the same as that on earth pressures, but once again there is a failure to generalize. Perhaps it is too much to expect that Cauchy's development of the stress tensor (in 1823) could have been anticipated, but Coulomb did make clear statements of equilibrium, pp. 4–5; pp. 43–4, and, had he cut a triangular element from the continuum, he might have made a start on the analysis of stress.

The Mohr–Coulomb criterion

Coulomb uses the problem of compressive fracture of columns to introduce the more difficult question of the thrust of soil against a retaining wall; further, his discussion of columns starts with the simplest case in which friction is neglected (article 8, p. 10; p. 48). It is convenient here to follow this same order in the presentation of the two-dimensional analysis of stress.

The sign convention to be adopted is illustrated in fig. 4.3(*a*). Reference axes *Ox*, *Oy* are taken, and it is required to compute the stresses σ_{aa}, σ_{bb} and τ_{ab} ($= \tau_{ba}$) referred to a general set of axes *Oa*, *Ob*. If σ_{xx} etc. are the corresponding values referred to the

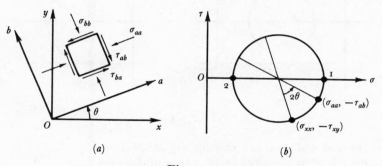

(*a*)   (*b*)

Fig. 4.3

original axes, then the quantities are related by the stress tensor, for which MOHR in 1882 devised the graphical representation of fig. 4.3(*b*); this work had been partly anticipated by CULMANN in 1866. Compressive stresses in fig. 4.3 are denoted *positive*. The points 1 and 2 in fig. 4.3(*b*) represent principal stresses, and correspond to planes (separated by $\frac{1}{2}\pi$) on which the shearing stress is zero.

Setting $\phi = 0$ in equation (4.1), and noting that the shear stress can act in either direction, Coulomb's criterion for failure of frictionless materials is simply $\tau = \pm c$; this is the maximum shear stress criterion of failure of perfectly plastic solids. This criterion defines a permissible region of the σ, τ plane within which all Mohr's circles must lie. Thus, in fig. 4.4(*a*), the circle shown (which represents the state of stress at a particular point in the continuum) can be of any radius less than *c*. Further, it can be centred at any point on the σ-axis; it is the *principal stress difference*, ($\sigma_1 - \sigma_2$) in fig. 4.4(*a*), which gives the diameter of the circle. The maximum shear stress criterion thus predicts that yield will occur when the maximum shear stress has value c, independently of the hydrostatic pressure $\frac{1}{2}(\sigma_1 + \sigma_2)$.

Thus if the column of fig. 4.1 is in a state of uniform stress, then failure under the compressive load P must be represented by a Mohr's circle of radius c, and it remains to locate this circle in the σ, τ plane. Now the principal directions 1 and 2 are immediately apparent in fig. 4.1; the horizontal top face of the column is not subjected to shear, but only to a (uniform) compressive stress, and the vertical

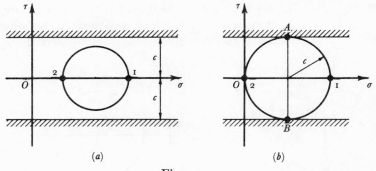

(a) (b)

Fig. 4.4

sides of the column are completely stress free. The principal axes therefore lie horizontally and vertically, and point 2 of the general Mohr's circle of fig. 4.3(b) must, in this particular problem, be located at the origin, fig. 4.4(b). From this last diagram it is seen at once that the compressive stress to cause fracture is given by $\sigma_p = \sigma_1 = 2c$. This result of course agrees with that of equation (4.9) on setting $\phi = 0$.

The failure planes are located by means of points A and B in fig. 4.4(b), which are distant $\frac{1}{2}\pi$ from the principal points 1 and 2 in the σ, τ plane. The directions of slide, along planes of maximum shear, are at right angles to the A and B directions, and are commonly labelled α and β. In the real body, therefore, the α and β lines along which failure occurs are located at $\frac{1}{4}\pi$ from the 1 and 2 directions, as shown in fig. 4.5. This angle of $45°$ was found directly by Coulomb, p. 11; p. 48, and accords with the general result of equation (4.8).

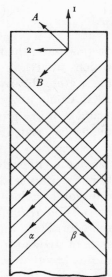

Fig. 4.5

The general failure criterion of equation (4.1) may now be examined. The lines

$$\pm\tau = c + \sigma\tan\phi$$

plot as shown in fig. 4.6, and again define a permissible region within which Mohr's stress circles must lie; the circle shown in the figure gives the limiting state for fracture of the column. Comparison with fig. 4.4(b) shows that points 1 and 2 representing the principal directions are located as before, but that points A and B, which determine

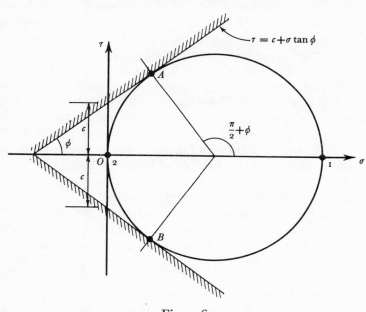

Fig. 4.6

the planes of failure, no longer represent planes of maximum shearing stress. Instead, point A (for example) is distant $(\frac{1}{2}\pi + \phi)$ from point 1 in the σ, τ plane, so that the A-direction in the real body makes an angle $(\frac{1}{4}\pi + \frac{1}{2}\phi)$ with the 1-direction; this is the result of equation (4.8). The α and β lines are in the directions sketched in fig. 4.7.

It is easily established from fig. 4.6 (for example, by similar triangles) that the radius of the circle shown is

$$\frac{c\cos\phi}{1 - \sin\phi} = c\tan(\tfrac{1}{4}\pi + \tfrac{1}{2}\phi) = c\cot s, \qquad (4.10)$$

so that the stress at point 1, which has just double the value (4.10),

115

gives the fracture stress of the column in conformity with equation (4.9).

The α and β directions along which slip occurs are, in the general case, the *characteristics* of the first-order partial differential equations of equilibrium of stress for a material in which Coulomb's criterion (4.1) is satisfied everywhere (see, for example, SOKOLOVSKII). Solutions to some problems of plasticity in the continuum can thus be obtained by examining statements of equilibrium and of the yield criterion, without reference to any ideas of compatibility of deformation. Both the fracture of columns and the thrust of soil against a retaining wall, as considered by Coulomb, are of this statically-determinate type. By ignoring compatibility conditions, lower-bound solutions are obtained to problems in plasticity.

However, Coulomb's treatment of the problem of the fracture of columns is in fact an upper-bound approach. His postulate of a plane of failure, whose location is unknown at the start of the analysis, is equivalent, in modern terms, to the specification of a velocity field, and equation (4.5) may be found directly by using a work equation rather than considering the balance of forces as in fig. 4.1.

Upper bounds from velocity fields

Considering first the frictionless ($\phi = 0$) case, the shearing stress acting on the assumed failure plane, fig. 4.8, has the constant value c. If the relative velocity of sliding is v, then the rate at which work is dissipated is

$$(ca^2\sec\theta)\, v, \tag{4.11}$$

where $a^2\sec\theta$ is the area of the fracture plane. The *vertical* velocity of the load P is $v\sin\theta$, so that the rate at which the external load does work is

$$Pv\sin\theta. \tag{4.12}$$

Equating expressions (4.11) and (4.12) leads to

$$P = \frac{ca^2}{\sin\theta\cos\theta}, \tag{4.13}$$

as found by Coulomb in article 8, p. 11; p. 48.

The value of P resulting from equation (4.13) for any assumed value of θ is an upper bound on the true value of the load required to cause collapse; this is an example of one of the fundamental theorems of plasticity (see e.g. PRAGER). Thus the value of P must be minimized to give the best upper bound, and, providing failure does indeed occur by sliding along a plane surface (i.e. that there is not some other class of mechanism which is more critical, such as sliding along a curved surface), then the least upper bound will in fact give the actual collapse load.

The frictional case may perhaps best be discussed by using first a more general yield criterion than (4.1). Suppose that failure occurs when the stresses τ and σ have values satisfying some given relation

$$f(\tau, \sigma) = \text{const.} \qquad (4.14)$$

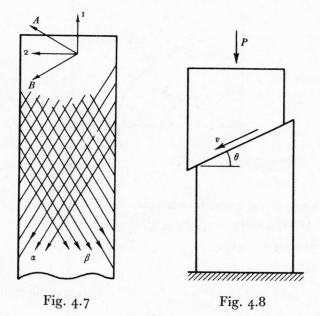

Fig. 4.7   Fig. 4.8

At a failure surface there is in general the possibility of displacement in *two* directions, corresponding to the two stresses acting on the surface; that is, in addition to a slide in the direction of the shearing stress τ (as in fig. 4.8), motion can also take place at right angles, in the direction of σ.

Thus, when a failure mechanism is postulated, two velocities, say v_τ and v_σ may be assigned. If now the problem is one which falls within the class which may be dealt with by the plastic theorems then the *normality condition* must be obeyed, that is

$$\frac{v_\tau}{v_\sigma} = \frac{\partial f/\partial \tau}{\partial f/\partial \sigma}. \qquad (4.15)$$

If the yield relation f of equation (4.14) is of the typical form sketched in fig. 4.9, and if deformation axes are superimposed on the stress axes, then equation (4.15) states that the vector representing the possible deformation at point P must lie normal to the yield curve at that point.

In solving a problem, a guess could be made for the mechanism of failure, that is, values of v_τ and v_σ might be postulated. Equations (4.14) and (4.15) could then be solved simultaneously to give the

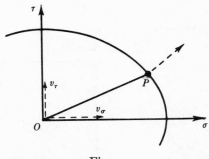

Fig. 4.9

values of τ and σ. As a specific example, if the curve in fig. 4.9 is the Mises (or Tresca) ellipse

$$\tau^2 + k\sigma^2 = c^2, \tag{4.16}$$

then equation (4.15) gives

$$\frac{v_\tau}{v_\sigma} = \frac{\tau}{k\sigma}. \tag{4.17}$$

Equations (4.16) and (4.17) solve to give the general coordinates of point P in fig. 4.9:

$$\left.\begin{array}{l} \tau = \dfrac{kcv_\tau}{[k(v_\sigma^2 + kv_\tau^2)]^{\frac{1}{2}}}, \\[3mm] \sigma = \dfrac{cv_\sigma}{[k(v_\sigma^2 + kv_\tau^2)]^{\frac{1}{2}}}, \end{array}\right\} \tag{4.18}$$

and, finally, the rate per unit area of dissipation of work at the failure surface is give by

$$\tau v_\tau + \sigma v_\sigma = c\left[v_\tau^2 + \frac{1}{k}v_\sigma^2\right]^{\frac{1}{2}}. \tag{4.19}$$

Expression (4.19) may be integrated over the failure surface and the result equated to the rate of working of the external load to give an upper bound on the collapse value of that load. The two velocities v_τ and v_σ may then be varied to give the best upper bound.

In attempting to use this technique with Coulomb's equation a severe restriction arises because of the linear form:

$$f(\tau, \sigma) \equiv \tau - \sigma \tan \phi = c. \tag{4.20}$$

Equation (4.15) indicates at once that

$$v_\sigma = -(\tan \phi)\, v_\tau; \tag{4.21}$$

that is, a sliding velocity v_τ *must* be accompanied by a *dilatant* velocity $v_\tau \tan \phi$, quite independently of the values of σ and τ, and arbitrary values can no longer be assigned to v_τ and v_σ. Interpreted

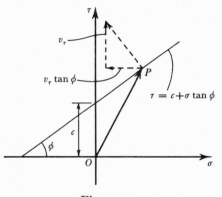

Fig. 4.10

geometrically, fig. 4.10, the normal to the yield curve at a general point P always has the same direction. Further, the rate of dissipation of work per unit area of the failure surface is

$$\tau v_\tau + \sigma v_\sigma = \tau v_\tau + \sigma(-v_\tau \tan \phi) = c v_\tau, \tag{4.22}$$

which is independent of the actual magnitudes of the stresses acting on the failure surface.

Thus, in assuming a mechanism for compressive fracture of the column, fig. 4.11, a sliding velocity v at the fracture plane will be accompanied by a velocity $v \tan \phi$ at right angles. From (4.22), the rate of dissipation of work at the fracture plane is still given by (4.11), but the external load P does work at the rate

$$P(v \sin \theta - v \tan \phi \cos \theta). \tag{4.23}$$

Equating (4.11) and (4.23) leads at once to the correct expression (4.5) for the value of P.

These ideas will be explored further in the next chapter which discusses the question of soil thrust against retaining walls.

In test 2 (Coulomb's fig. 2) it is not clear how nearly a state of pure shear was achieved. It is fairly certain that some bending must have been present due to imperfect fit in the mortise, and this makes any numerical discussion of the test results difficult. However, Coulomb was convinced that the limiting cohesive stresses in pure shear and pure tension had virtually the same value (p. 7; p. 45); in his computations for brick on p. 13; p. 50 he uses for c the value deduced from a tension test.

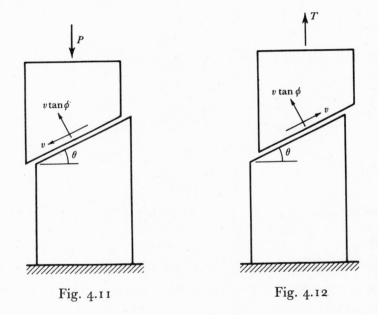

Fig. 4.11   Fig. 4.12

Now for a material having properties (c, ϕ) and which fails according to 'Coulomb's equation' (4.1), it is as easy to do a tensile as a compressive analysis. The proper failure mechanism for a column in tension is shown in fig. 4.12 (cf. fig. 4.11); note the same phenomenon of separation as well as sliding of the two portions of the column. The work equation gives

$$T = \frac{ca^2}{\cos\theta(\sin\theta + \tan\phi\cos\theta)}, \qquad (4.24)$$

120

which may be compared with equation (4.5). The minimum value of T occurs for

$$\theta = (\tfrac{1}{4}\pi - \tfrac{1}{2}\phi) = s, \tag{4.25}$$

which therefore defines the direction of the characteristics; the failure load has value

$$T_{\text{min}} = 2ca^2 \tan s. \tag{4.26}$$

These same results could have been obtained from a plot of Mohr's circle, fig. 4.13. Comparing with fig. 4.6, the point 2 is again located at the origin, but the circle lies wholly to the left of the τ-axis, imply-

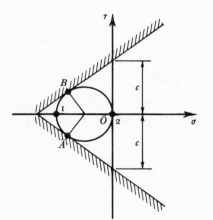

Fig. 4.13

ing tensile stresses throughout. The circle is the largest that can be drawn, and equations (4.25) and (4.26) can be established from the geometry of the diagram.

It is a remarkable coincidence that for $\tan \phi = \tfrac{3}{4}$, the value of $\tan s \ (= \tan (\tfrac{1}{4}\pi - \tfrac{1}{2}\phi))$ is exactly equal to $\tfrac{1}{2}$, so that $T_{\text{min}} = ca^2$ from equation (4.26); the apparent tensile cohesive strength will then exactly equal the shearing cohesive strength of the column. This coefficient of friction of $\tfrac{3}{4}$ was precisely that used by Coulomb (article 4, p. 5; p. 44).

(Assuming indeed that $\phi = \tan^{-1}\tfrac{3}{4}$ for brick, then, from equation (4.25), the failure plane would be inclined at 26° 34′; this angle is the complement of Coulomb's 63° 26′, p. 13; p. 49. Coulomb does not describe the tensile fracture. If the surface was in fact inclined, its actual location could have been arbitrary, although an edge of the fracture would have been likely to lie at one end of the notch, such as *ef* in Coulomb's fig. 1.)

121

It is, of course, a major assumption that the material parameters c and ϕ can be assigned constant values; for example, BOUASSE in 1901 pointed out that not only could they be functions of say temperature, but they could depend on the ambient values of the stresses. Ten years later KÁRMÁN reported tests on marble and sandstone, in which the specimens were compressed axially while at the same time being subjected to a hydrostatic pressure which could be varied from zero up to several thousand atmospheres. He found indeed that the apparent values of c and ϕ were strongly dependent

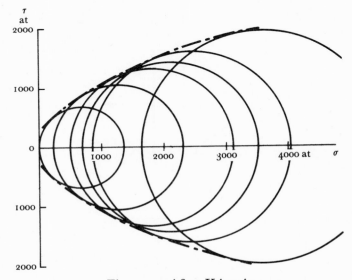

Fig. 4.14. After Kármán.

on the hydrostatic pressure; the value of ϕ for sandstone, for example, dropped from about 50° at low pressures to about 20° for high pressures. Figure 4.14 reproduces Kármán's fig. 7 for marble.

However, these results are right outside any practical range. Mean stresses in masonry construction might lie in the range 5–50 atmospheres (1 atm. $\sim$ 100 kN/m²) and the unconfined crushing strength of a medium sandstone is about 500 atmospheres. Kármán's tests extended up to a hydrostatic pressure of almost 10 times this unconfined strength. Thus the results would have had no direct interest for Coulomb, and have no relevance to his work; he wished 'to determine...the effect of friction and of cohesion' in problems which had arisen in his own experience.

His concern was, in fact, to establish a *linear* and *empirical* theory which involved *two* material constants (c and ϕ), and he is quite clear that the material constants in his equation (4.1) are to be established by experiment. Thus for the friction angle ϕ 'it will be well to make tests on the materials to be used' (p. 5; p. 44); his own value of $\frac{3}{4}$ is introduced only by way of illustration. For the cohesion of mortar (p. 7; p. 45), 'individual tests must be made in each place'. The fact that a tension test happened by accident to give a reasonable value of c merely perhaps saved Coulomb some experimental work; his test 2 *was* an attempt to measure c directly.

Further, having determined the values of c and ϕ, Coulomb manipulates the linear yield criterion completely correctly, and he gives in the Introduction, pp. 1–2; pp. 41–2, a clear exposition of the upper-bound theorem of plastic theory. Finally, the linear theory of equation (4.1) explained satisfactorily the experimental data known to Coulomb.

5

The Thrust of Soil

GAUTIER in 1717 listed 'cinq Difficultez proposées aux Sçavans, à résoudre', namely

1. the thickness of abutment piers for all kinds of bridges;
2. the dimensions of internal piers as a proportion of the span of the arches;
3. the thickness of the voussoirs between intrados and extrados in the neighbourhood of the keystone;
4. the shape of arches;
5. the dimensions of retaining walls to hold back soil.

He states that he will use 'un peu de Physique, de raisonnement & d'experience', together with 'les Méchaniques' and 'quelque chose de la Statique' to satisfy the equilibrium of forces, as well as 'la Geometrie' (i.e. the mathematics of curves and surfaces); and he hopes 'que le moindre Ouvrier avec le sens commun, pourra tracer & démontrer ce que j'avance'. All this is reminiscent of Coulomb's 'mélange du Calcul & de la Physique' in his opening sentence, and of his concern (p. 4; p. 43) that a workman with little learning should understand his principles.

Gautier's fifth problem, that of the thrust of soil, was to become one of the 'classic' problems of the eighteenth century, and was relatively new; the solution was required in order to proportion the retaining walls for the deep cuttings in Vauban's system of defence. In the previous century, BLONDEL's 'Resolution des quatres principaux problemes d'architecture' does not mention soil thrust; his four problems are concerned with the entasis of columns, the shape and the jointing of arches, and the shape of (cantilever) beams of greatest strength. Thus Coulomb's four topics (the strength of beams, the strength of columns, the thrust of soil, and the thrust of arches) had been current for about a hundred years, or just less in the case of soil.

MAYNIEL believes that BULLET, in his book of 1691, 'was the first to try to establish a theory of earth pressure based on the principles of mechanics' (Mayniel's book of 1808 will be discussed more fully later). Bullet's *L'architecture pratique* gives rules and quantities for building, and deals with masonry, woodwork (*charpenterie*) and so on. On p. 159 starts the section 'De la Construction des murs de Rempart & de Tarrasse', and Bullet states that there are three things to note in the construction of walls in general: first, the quality of the materials and how they are employed; second, the quality of the earth so that their foundations may be well laid; third, their thickness and batter.

On the first two topics Bullet is merely descriptive and empirical, but his remarks on the proper sorts of earth to build on are of interest. He recommends the making of a large number of trial holes 'trous ...en forme de puits') to find the different strata ('differens lits de terre') in order to make sure that an apparently good soil does not overlay a clay, or a sandy soil, or some other soil which can be compressed by a load. If the trial holes cannot be made, then the earth may be beaten with a wooden rafter six or eight feet long; if the sound is dry and light, and the soil offers resistance, then the earth is firm, but a heavy sound and poor resistance mean a worthless foundation.

Marshy land will require piled foundations, and Bullet gives the rule that the ratio length to diameter of the piles should be about 12. ('Cette regle est selon les bons Auteurs.') However, Bullet himself thinks that for piles longer than 12 ft, smaller diameters may be used; he gives 13 or 14 in for lengths of 16 or 18 ft. (It may be noted that if the Euler buckling load is set equal to the squash load, then $\pi^2 EI/L^2 = \sigma_0 \pi D^2/4$, from which $L/D = \frac{1}{4}\pi\sqrt{(E/\sigma_0)}$; setting $E/\sigma_0 = 200$, which is a reasonable value for oak, $L/D = 11$.) Bullet continues:

The third thing that must be noted in the building of ramparts and terrace walls is to know how to give them an appropriate thickness in proportion to the height of soil they must retain...This rule has not yet been given by any writer on Architecture, whether civil or military.

Bullet then presents 'un essay...qui est fondé sur les principes de Mechanique'.

He starts by noting that sand is 'la terre la plus coulante'. If sand is supposed to be made up of small round grains, arranged in a 'natural disposition', fig. 5.1, then the natural slope is seen to be 60°. It would seem that a sandy soil could not have a slope steeper than this, but in practice some do. However, for the sake of safety, Bullet proposes to take a natural slope of 45° in his calculations. This figure of 45° recurs again and again in subsequent work on earth pressure, and is, for example, used by Coulomb in his numerical example on p. 19; p. 54; it has already been noted that Bélidor used the same angle, and this was probably Coulomb's immediate source.

Thus if a soil is cut vertically as at *AB*, fig. 5.2, and retained by a wall, and this wall is then suddenly removed, the earth will fall to form the inclined surface *BC*, this corresponding to the assumed natural angle of slip (*écoulement*); what is needed is the calculation of the force required to balance the thrust of the wedge (*coin*) *CAB*.

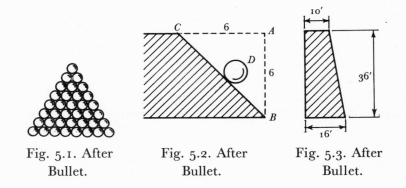

| Fig. 5.1. After | Fig. 5.2. After | Fig. 5.3. After |
| Bullet. | Bullet. | Bullet. |

Bullet states that the (inclined) force necessary to hold a ball on the slope *BC* has value $1/\sqrt{2}$ (or 5/7) of the weight of the ball, so that, by extension to the whole mass, a force equal to 5/7 of the weight of the triangle *CAB* must be resisted by the wall. Numerically, if $AB = AC \doteq 6$ toises, then the force will be proportional to

$$\tfrac{5}{7}(18) = 13 \text{ toises}^2,$$

so that the retaining wall (assumed to be of the same density as the soil) must have this same area. Thus, since 1 toise = 6 ft, the wall (with a batter of 1 in 6) will have the dimensions of fig. 5.3.

Gautier's book quotes Vauban's rules for determining the profiles of retaining walls, and refers to Bullet's theory and results without, however, attempting to correct Bullet's rather arbitrary statics. Gautier's table of profiles, for heights from 5 ft to 80 ft in steps of 5 ft, with a batter of 1 in 5, gives dimensions considerably slimmer than those of Bullet; for example a 40 ft wall (cf. fig. 5.3) has a ridge width of 4 ft 4 in and a base width of 12 ft 4 in.

COUPLET presented two *Mémoires* on soil thrusts to the *Académie*, in 1726 and 1727, to which he added a third in 1728; this last, however, merely uses the ideas of the first two to calculate the dimensions of retaining walls with added supporting buttresses. (In the two following years, 1729 and 1730, he presented two further remarkable papers on the thrust of arches; these are discussed in chapter 6. Thus Couplet like Coulomb fifty years later, was working on two of the major problems of civil engineering.)

Couplet's first paper on earth pressure starts by noting that both Bullet and Gautier made considerable errors both in the calculation of forces and in the way of considering the natural slope of the soil. First, Bullet's slope of 60° is wrong (fig. 5.1); this is a two-dimensional

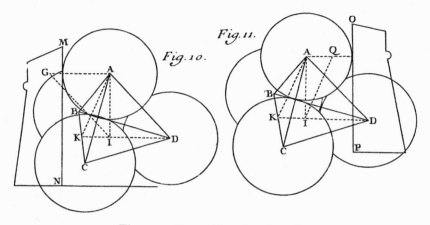

Fig. 5.4. From Couplet (1726).

configuration, whereas in reality four balls will be arranged as a tetrahedron (fig. 5.4, Couplet's figs. 10 and 11). Second, Bullet's factor of $1/\sqrt{2}$ does not give the *horizontal* thrust of the soil. Third, the natural slope of a soil cannot be considered as an inclined plane down which the retained wedge of soil is trying to slide.

To deal with this last point, Couplet supposes that in Bullet's 60° configuration, fig. 5.5 (Couplet's fig. 7), one ball D is outside the natural slope CB. This ball would then slide not on CB, but on an imaginary plane LK, and the force necessary to hold it in place will depend on the slope of LK. By contrast, he shows correctly that the smooth wedge theory implies a *constant* horizontal thrust against the (smooth) retaining wall independent of the angle of repose, and that the thrust is simply proportional to $\frac{1}{2}h^2$, where h is the height of the retaining wall.

127

Couplet applies these ideas to his tetrahedral packing, fig. 5.4, and has some difficulty in deciding whether AK in his fig. 10 or AD in his fig. 11 represents the natural slope. He does not resolve this point, but shows that in either case the thrust of the soil against a smooth wall will be the same. He arrives at an overturning force proportional to $\frac{1}{8}h^2$ acting at a height $\frac{2}{3}h$ above the base of the wall, giving an overturning moment of $\frac{1}{12}h^3$. This is Couplet's basic result in this first *Mémoire*; if, for example, a rectangular retaining wall has thickness x and density π, the soil density being γ, then

$$\tfrac{1}{2}\pi h x^2 = \tfrac{1}{12}\gamma h^3 \quad \text{or} \quad x = h\left(\frac{\gamma}{6\pi}\right)^{\frac{1}{2}}.$$

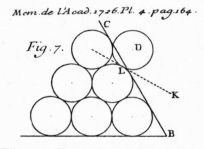

Fig. 5.5. From Couplet (1726).

(For $\gamma = \pi = 1$, and $h = 36$ ft, then $x = 14$ ft 8 in, which is little different from Bullet's dimensions, fig. 5.3.) Couplet applies his basic result to ten different profiles of wall, and shows how to calculate the dimensions of each.

In his second *Mémoire*, Couplet takes into account the effect of friction of the soil against the retaining wall. This is necessary, he says, because the results of the frictionless theory are notably at variance with the dimensions for retaining walls established through experience by 'nos plus habiles Ingénieurs & Architectes'. The effect of wall friction is, according to Couplet, to allow the soil thrust to act in an inclined direction and no longer horizontally, so that the overturning moment on the wall is considerably reduced. He is still unable to decide whether AK or AD in fig. 5.4 should be considered as the natural slope, and indeed he adds a third packing arrangement, that of the square-based pyramid, which gives him a greater overturning moment than the other two. He presents his calculations in the form of three tables, one for each of the three assumptions of natural slope; the profiles are very much slimmer than those of his first *Mémoire*, and more like those of Gautier.

As has been noted, B É L I D O R reverted firmly to the idea of a sliding wedge (of angle 45°) in 1729, and was really just as successful, in a practical sense, as Couplet in establishing tables of wall profiles. Couplet was attempting a 'scientific' analysis, and Bélidor was more frankly empirical, but both agreed (with Gautier and Bullet) that the thrust of soil against a retaining wall was proportional to the square of the height of the wall, and that the overturning moment was proportional to the cube of the height; what was being attempted was the calculation of the constants in these formulae. (Bélidor in fact, arrived at the same basic result for overturning moment as Couplet: $\frac{1}{12}\gamma h^3$.)

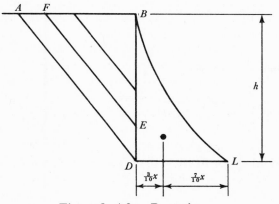

Fig. 5.6. After Papacino.

In this connection the proposals of P A P A C I N O D ' A N T O N I in book 5 of his *Architettura militare* are of interest. (This volume was published in 1781 for the use of the school of artillery at Turin, but must have been in existence in manuscript for a decade or so earlier; biographical details of Papacino are given by Balbo.) Arguing from similar triangles, Papacino concludes that since both *AD* and *FE* in fig. 5.6 are failure planes, the thrust on the wall must be proportional to the square of the height, and since this must be true for any height, the best shape of wall is parabolic. An easy integration gives the resisting moment of the wall to overturning about the toe *L* as $\frac{3}{70}\pi hx^2 + \frac{1}{2}tx^2$, where π is the unit weight of the wall and t is the *adesione* resisting separation along *LD*. This expression may be equated to the overturning moment (which Papacino takes, in conformity with Couplet's basic result, to be $\frac{1}{12}\gamma h^3$) to give the required dimension x.

However, these calculations do not really lead anywhere. Papacino himself advances reasons for not building in practice a wall of parabolic profile; among other things, he is worried about the run-off of rainwater and, indeed, advocates that a straight batter should never exceed 1 in 5. Further, his subsequent calculations, including walls both with batter and with buttresses, specifically neglect the cohesion of the masonry.

Papacino d'Antoni does not make the assumption of a natural slope of 45° (although in fact, on the smooth wedge theory, the angle of repose is irrelevant, as Couplet has shown, and Papacino's basic result of $\frac{1}{12}\gamma h^3$ does not involve this angle). Instead, he notes that dry sand might have an angle of repose of about 30°, while *terra grassa* could have an angle of 50°; if the soil also has *tenacità*, then the angle could be 65°. These angles are similar to those quoted by QUERLONDE in a *Mémoire* of 1743, listed and discussed by MAYNIEL in 1808. (Mayniel discusses several *Mémoires* which were apparently in manuscript and which appear never to have been published; for example, the *Mémoire* by SALLONNYER of 1767 mentioned below was found by Mayniel in the *dépôt des fortifications*.) Querlonde gives an angle of $\tan^{-1}\frac{1}{2}$ (about 27°) for *terres sabloneuses*, 45° for *terres végétales*, and $\tan^{-1}\frac{3}{2}$ (about 56°) for *terres fortes*.

(These numbers contrast with Bélidor's *principe d'expérience*, which defends his choice of 45° while allowing that different soils could have different slopes:

It is known from experience that ordinary soils, when they are newly-turned and placed without ramming and not interlaced with faggots, take a slope or batter which makes an angle of 45° with the horizontal, and this I say is the case for ordinary soils...

(while for sandy soil the angle is more acute, but steeper for clayey soil).)

Of all these writers, Sallonnyer seems to be the only one who proposes an overturning moment not directly proportional to the cube of the height of the wall. By resolving forces (and Mayniel comments that Sallonnyer was the first to allow for the decomposition of forces, that is, to examine the statical equilibrium of the sliding wedge), he obtains both horizontal and vertical components acting on the back of the wall. To understand the derivation of the vertical component, it is necessary to go back to Bélidor's analysis. It will be remembered (chapter 2) that Bélidor had computed the thrust as that corresponding to the fluid pressure $\frac{1}{2}\gamma h^2$, but somewhat arbitrarily reduced by a factor $\frac{1}{2}$ to the value $\frac{1}{4}\gamma h^2$ to allow for the *tenacité* of the soil; the thrust acts at a height $\frac{1}{3}h$ above the foot of the wall. It is clear that Bélidor imagines some frictional action on the plane of sliding, so that his system of forces (which he did not in fact analyse) is as shown in fig. 5.7. Sallonnyer appears to have retained both the magnitude and line of action of Bélidor's thrust, but he supposes that the wedge is sliding without friction on the plane of failure. An extra vertical force is then necessary for equilibrium, and this must appear on the back face of the wall, fig. 5.7.

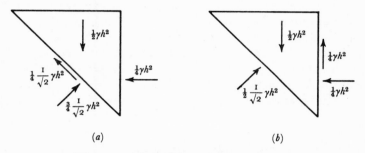

Fig. 5.7. (*a*) Bélidor; (*b*) Sallonnyer.

The horizontal component gives the familiar quantity $\frac{1}{12}\gamma h^3$, from which must be subtracted, however, $\frac{1}{4}\gamma h^2 x$ arising from the vertical force, where x is the thickness of the (rectangular) retaining wall. Thus, for stability,

$$\pi(\tfrac{1}{2}hx^2) \;>\; \gamma(\tfrac{1}{12}h^3 - \tfrac{1}{4}h^2 x).$$

It is of interest that Sallonnyer's system of forces implies an angle of friction between the soil and the rough retaining wall at least equal to the internal angle of friction of the soil, $45°$.

This *Mémoire* of Sallonnyer (1767) would probably have been unknown to Coulomb; the two terms in the value of the overturning moment have a fortuitous resemblance to those of Coulomb's formula $\frac{1}{3}mh^3 - \frac{1}{2}\delta lh^2$ (p. 19; p. 54), but the derivation is completely different. Sallonnyer's theory was built directly on previous work, while Coulomb made a more radical departure; however, neither writer seems to have been much concerned with the results of actual tests. Even Bélidor, with his *principe d'expérience*, does not say that he made any experiments.

The first real tests on soils appear to be those of GADROY in 1746, as reported by Mayniel. Gadroy made a box 3 in square and 11 in long, of which one square end face (the retaining wall) was hinged and could be released. The box was filled with fine dry sand, and the end face dropped; Gadroy found that the initial plane of rupture did *not* coincide with the natural slope, but was steeper ($\tan^{-1}\frac{3}{2}$). However, he did not seem to find this significant, and again it is in any case unlikely that Coulomb would have seen the work. The only other reported experiments before 1773 seem to be those of RONDELET (about 1767), but these, although made with a much larger box (12 in wide and $17\frac{1}{2}$ in high), added little to the work of Gadroy.

Thus Coulomb would have written the *Essai* with very few practical results to guide him. Although he worked in general terms with an arbitrary angle of friction, it is not surprising that he used 45° in his examples. Similarly, he is very ready to set the cohesion equal to zero when making practical calculations. However, it is of course precisely the introduction of *two* parameters c and ϕ describing the soil properties that is of the utmost importance in Coulomb's analysis. Bossut had considered the cohesion of the material of the *dyke* (equation (2.1)), and Papacino the *adesione* of the wall, but no writer before Coulomb seems to have introduced the possible cohesion of the *soil* into his theory. At best, Bélidor's arbitrary factor of $\frac{1}{2}$ on account of *tenacité* is an allowance for friction, and not for cohesion.

Coulomb's problem (1); $c = 0$

It has been noted that what is now known as Coulomb's equation, equation (4.1), was not written in that form in the *Essai*; on the other hand, his basic result for the value of the thrust of soil against a retaining wall,

$$A = mh^2 - \delta lh \tag{5.1}$$

(p. 18; p. 53), may well be named after him. The thrust is a linear combination of two terms, of which the first is the familiar 'fluid pressure', for which the coefficient m, however, is no longer an arbitrarily assumed factor, but an analytically derived function of the angle of internal friction. From this first term, which is proportional to the square of the height of the wall, is subtracted a second directly proportional to the height, and whose coefficient involves both internal friction and cohesion. It is instructive to derive equation (5.1) using the methods of upper and lower bounds that were outlined in chapter 4, and, further, to discuss first the case of a cohesionless soil for which $c = 0$.

Coulomb makes the enormous advance of not assuming *a priori* the angle of slip of the soil at incipient failure; in particular, he does not equate this angle to the angle of natural repose. The analysis follows very closely that given in chapter 4. In fig. 5.8 Coulomb's parameter x has been abandoned in favour of an unknown angle α defining the slip plane; simple resolution of forces (as in note 19) gives

$$\left. \begin{aligned} P &= A\cos\alpha + W\sin\alpha, \\ Q &= -A\sin\alpha + W\cos\alpha. \end{aligned} \right\} \tag{5.2}$$

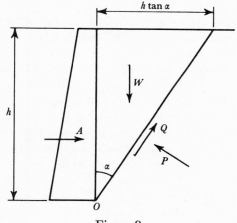

Fig. 5.8

In writing equations (5.2), it has been assumed that the force A acts horizontally, that is, that there is no friction between the earth and the retaining wall; setting $Q = P\tan\phi$,

$$A = W\cot(\alpha + \phi), \tag{5.3}$$

and noting that

$$W = \tfrac{1}{2}\gamma h^2 \tan\alpha, \tag{5.4}$$

where γ is the unit weight of the soil, then

$$A = \tfrac{1}{2}\gamma h^2 \tan\alpha \cot(\alpha + \phi). \tag{5.5}$$

This value of A is a maximum when

$$\alpha = (\tfrac{1}{4}\pi - \tfrac{1}{2}\phi) \equiv s, \tag{5.6}$$

the corresponding value being

$$A_{\max} = \tfrac{1}{2}\gamma h^2 \tan^2 s. \tag{5.7}$$

Equation (5.5) may also be found by considering the wedge to slide along the plane of fracture, due allowance being made for

133

dilatation. Thus in fig. 5.9 (cf. fig. 4.11) a sliding velocity v must be accompanied by a velocity $v\tan\phi$ at right angles; from equation (4.22), the rate of dissipation of work at the fracture plane is zero, since the present analysis assumes that the cohesion c is zero. Thus only the forces A and W in fig. 5.9 do work, and

$$A(v\sin\alpha + v\tan\phi\cos\alpha) = W(v\cos\alpha - v\tan\phi\sin\alpha),$$

or
$$A = W\cot(\alpha+\phi), \quad \text{as before.} \tag{5.3 bis}$$

The maximum value of A given by equation (5.7), where s has the value (5.6), is a *relative* maximum; there might be some other *class* of mechanism (e.g. with curved failure surfaces, as envisaged by Coulomb) which gives a worse case. Ignoring this point for the moment, and confining attention to the triangular wedge of failure, the value of A given by equation (5.7) is not, in itself, sufficient information for the design of the retaining wall; the height at which A acts must also be known. To determine this height, Coulomb's own method may be followed; alternatively, the value of the over-turning moment may be found directly by considering a small virtual rotation of the retaining wall about O in fig. 5.8.

Assuming, with Coulomb, that the *whole* of the triangular wedge is in the critical state, then failure will occur by rupture along the characteristics, of which typical lines are sketched in fig. 5.10(a). If then the retaining wall is given a virtual angular velocity $\frac{1}{2}\omega$ as in fig. 5.10 (b) (the reason for the value $\frac{1}{2}\omega$ rather than ω will become apparent), the wedge of soil will deform to the state shown, typical velocities at a general point in the soil being given by the components (u, v). The actual values of u and v are given at once by consideration of the only permitted deformations along the rupture lines. The 'lozenge' marked in fig. 5.10 (a) is shown enlarged in fig. 5.10 (c); the permitted motion may be described by a simple four-bar chain mechanism with links bc and ad having angular velocity ω, while links ab and dc have zero angular velocity. The velocity vector at c will then make an angle ϕ with the fixed direction cd (or ab), and this will be true of all points in the wedge; the shearing velocity is thus accompanied by the correct dilatant velocity, as in fig. 5.9. For a cohesionless soil, no work will be dissipated in such a velocity field.

The values of u and v corresponding to the motion of fig. 5.10 (c) are given by

$$u = \tfrac{1}{2}\omega(y - x\cot s),$$

and
$$v = \tfrac{1}{2}\omega\tan s(y - x\cot s). \tag{5.8}$$

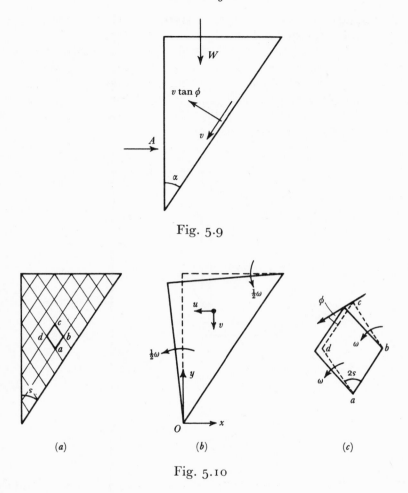

Fig. 5.9

Fig. 5.10

It will be seen that u and v are both zero along the sloping side of the wedge, and that both the retaining wall and the top surface of the soil have angular velocities $\frac{1}{2}\omega$. At the centre of gravity of the wedge $(x = \frac{1}{3}h\tan s, \; y = \frac{2}{3}h)$, the value of v is $\frac{1}{6}\omega h\tan s$, and the virtual work done by the soil will be this expression multiplied by the weight of the wedge, $\frac{1}{2}\gamma h^2\tan s$. Thus, if the moment acting on the retaining wall is M, the balance of virtual work gives

$$M(\tfrac{1}{2}\omega) = (\tfrac{1}{2}\gamma h^2\tan s)(\tfrac{1}{6}\omega h\tan s),$$

or
$$M = \tfrac{1}{6}\gamma h^3\tan^2 s. \tag{5.9}$$

Comparing equations (5.7) and (5.9), the line of action of the force A is seen to be distant $\frac{1}{3}h$ above the base of the wall.

The soil in fig. 5.10 is in an *active Rankine state*, so-called following RANKINE's study of 1857, in which he examined the equilibrium of stress in a cohesionless granular medium; he established the characteristics sketched in fig. 5.10 (a) as planes of failure. In fig. 5.11 an element of soil at depth y below the free surface is subjected to a vertical compressive stress γy, and to a horizontal stress σ_x (at the

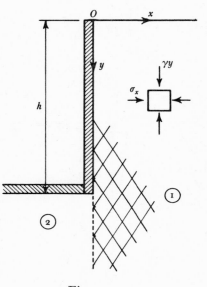

Fig. 5.11

moment unknown). Mohr's circle may be drawn as in fig. 5.12; if the circle just touches the yield envelope $|\tau| = \sigma \tan \phi$, then, immediately,

$$\sigma_x = (\gamma y) \tan^2 s, \qquad (5.10)$$

where (as usual) the value of s is given by equation (5.6). Thus the lateral pressure on the wall is seen to be proportional to the depth of soil; integration of equation (5.10) leads to the value A of thrust given by equation (5.7), and a second integration will give the value M of the overturning moment, equation (5.9).

The stress field of fig. 5.11 can be continued for all $x > 0$ and all $y > 0$, that is, for zone 1 in the figure. Provided some barrier is placed on the surface of the soil at $y = h$ for $x < 0$, capable of taking an upthrust, then it is possible to construct a stress field for zone 2 for which the Mohr's circles also lie within the permitted region of fig. 5.12; this extension of stress fields to cover the whole of the area will be discussed more fully below. (Crudely, in this case, the medium (sand) must be prevented from seeping out below the bottom of the retaining wall.)

Thus fig. 5.10(a) represents a velocity field which leads to an unsafe estimate of the value of the thrust A, while fig. 5.11, with the same characteristics, represents a stress field leading to a safe value

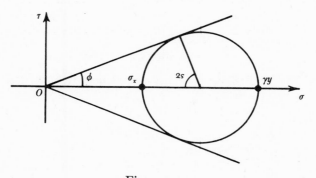

Fig. 5.12

of A. Since these two values of A in fact coincide, being equal to $\frac{1}{2}\gamma h^2 \tan^2 s$, the correct answer to this particular problem has been obtained.

It may be repeated that this correct solution applies to a *cohesionless* soil (e.g. dry sand) acting on a *vertical frictionless* retaining wall. Subsequent sections of this chapter will examine more complex cases of the problem.

Coulomb's problem (2); (c, φ)

Coulomb's main problem may now be examined, namely the evaluation of the thrust of a cohesive frictional soil against a smooth vertical retaining wall, leading to his two-term formula, equation (5.1); as will be seen, this formula does not give the correct solution to the problem.

Coulomb's own solution postulates a plane of failure, and the forces acting on the plane are related by 'Coulomb's equation'. Thus, in fig. 5.8, the forces P and Q (whose values are given by equations (5.2)) may be regarded as multiples of the stress σ and τ, and related by equation (4.1) $(\tau = c + \sigma\tan\phi)$:

$$(-A\sin\alpha + W\cos\alpha) = c(h\sec\alpha) + (A\cos\alpha + W\sin\alpha)\tan\phi,$$

that is
$$A = W\cot(\alpha+\phi) - ch\sec\alpha\cos\phi\operatorname{cosec}(\alpha+\phi)$$
$$= \tfrac{1}{2}\gamma h^2\tan\alpha\cot(\alpha+\phi) - ch\{\tan\alpha + \cot(\alpha+\phi)\}$$

or
$$A = (\tfrac{1}{2}\gamma h^2 + ch\cot\phi)\tan\alpha\cot(\alpha+\phi) - ch\cot\phi. \tag{5.11}$$

(The manipulation of this equation closely follows Prony's treatment.) Thus the condition for A to be maximum is exactly the same as when cohesion is neglected; cf. equation (5.5). The value of α is equal to s, equation (5.6), so that the characteristics have the same directions as in the cohesionless case, and the corresponding maximum value of A is
$$A = \tfrac{1}{2}\gamma h^2\tan^2 s - 2ch\tan s. \tag{5.12}$$

Coulomb's constants, equation (5.1), are thus determined as

$$\left.\begin{aligned} m &= \tfrac{1}{2}\gamma\tan^2 s, \\ l &= 2\tan s. \end{aligned}\right\} \tag{5.13}$$

Exactly as before, equation (5.11) can also be derived by considering the work balance for a virtual displacement of the wedge of failure. Corresponding to the displacements of fig. 5.9, the rate of dissipation of work at the failure plane is simply cv per unit length (equation (4.22)), so that

$$A(v\sin\alpha + v\tan\phi\cos\alpha) = W(v\cos\alpha - v\tan\phi\sin\alpha) - cv(h\sec\alpha).$$

Similarly, on the assumption that the *whole* of the wedge is in the critical state, fig. 5.10, the overturning moment on the wall may be found. The rate of dissipation of work in an elementary lozenge, fig. 5.10(c), is merely $c\omega$ times the area of the lozenge, so that, corresponding to equation (5.9),

$$M(\tfrac{1}{2}\omega) = (\tfrac{1}{2}\gamma h^2\tan s)(\tfrac{1}{6}\omega h\tan s) - (c\omega)(\tfrac{1}{2}h^2\tan s),$$

or
$$M = \tfrac{1}{6}\gamma h^3\tan^2 s - ch^2\tan s. \tag{5.14}$$

The Rankine analysis confirms equations (5.12) and (5.14), and at the same time shows that these equations, which correspond to those given by Coulomb, do not solve the problem posed. A general Mohr's circle for a portion of the soil in the critical state is shown in fig. 5.13; the general relation between the principal stresses at points 1 and 2 is
$$\sigma_2 = \sigma_1\tan^2 s - 2c\tan s. \tag{5.15}$$

Thus, from fig. 5.11, setting $\sigma_1 = \gamma y$,

$$\sigma_x = \gamma y \tan^2 s - 2c \tan s, \tag{5.16}$$

and a single integration leads to equation (5.12), while a further integration leads to equation (5.14).

Coulomb proposes (pp. 20–21; p. 55) an ingenious formula to determine the maximum height of an unsupported vertical cut. Setting $A = 0$ in equation (5.1) or equation (5.12) gives

$$h = c\frac{l}{m} = 4\frac{c}{\gamma}\cot s. \tag{5.17}$$

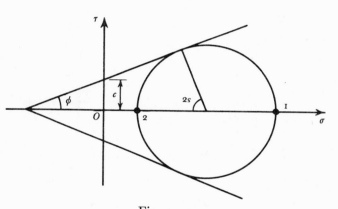

Fig. 5.13

However, a contradictory result is obtained if the overturning moment M in equation (5.14) is set equal to zero, rather than the overturning force A:

$$h = 6\frac{c}{\gamma}\cot s. \tag{5.18}$$

This contradiction is resolved if an examination is made of the distribution of the horizontal stress σ_x, equation (5.16). Figure 5.14 shows that the supposedly free edge of the cut is in fact subjected to a set of normal stresses, varying from tension at the top, through zero at a depth $2c\gamma^{-1}\cot s$, to compression at greater depths. The height given by equation (5.17) does correspond to zero *total* force on the vertical face of the cut, but this force must be distributed according to fig. 5.14, and hence the solution does not correspond to the actual problem.

Similarly, for the problem of the retaining wall, the value of A given by equation (5.12) implies that the soil is somehow 'glued' to the wall, so that the tensile stresses of equation (5.16), which occur for small values of y, can be developed. This again does not represent the actual situation and, moreover, by a theorem of DRUCKER (1954) which will be mentioned again later, the corresponding value of A is unsafe. Thus a greater overturning force than that given by equation (5.12) will actually act on the wall.

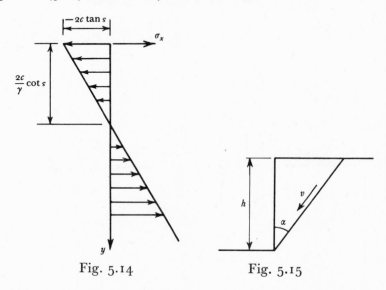

Fig. 5.14   Fig. 5.15

Before discussing further this problem of the retaining wall, it is instructive to examine more deeply Coulomb's subsidiary problem, that of the greatest depth to which a trench can be dug in horizontal soil.

The stability of a vertical cut (1); $\phi = 0$

As for the case of the fracture of a column, chapter 4, it is convenient to present the analysis first for a purely cohesive material, that is, to examine the perfectly plastic problem. In fig. 5.15, a trench has been dug to the critical depth h, so that the soil is on the point of falling into the cutting. Coulomb's wedge is shown making an angle α with the vertical, and a work balance gives

$$c(h \sec \alpha) v = (\tfrac{1}{2}\gamma h^2 \tan \alpha)(v \cos \alpha),$$

or

$$h = \frac{4c}{\gamma \sin 2\alpha}. \tag{5.19}$$

140

It is readily seen that, within the framework of the theorems of plasticity, the value of h given by equation (5.19) is an upper bound to the true value; in other words, the solution of equation (5.19) corresponds to a postulated mechanism of failure, and hence is unsafe. The value of h must be minimized, that is $\alpha = \frac{1}{4}\pi$ and $h = 4c/\gamma$ (in conformity with equation (5.17)).

Once again, this value of h is only a *relative* minimum; there might exist (in fact there does exist) another class of mechanism which is more critical. Thus, from equation (5.19),

$$h \leqslant 4c/\gamma. \tag{5.20}$$

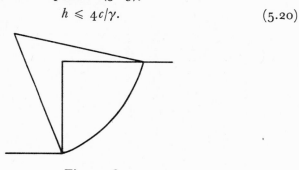

Fig. 5.16

The class of circular failure mechanisms of fig. 5.16 (see for example TAYLOR, or CHEN, GIGER and FANG) leads to the upper bound:

$$h \leqslant 3.83c/\gamma; \tag{5.21}$$

thus the plane failure mechanism is certainly incorrect. However, the circle mechanism is also incorrect, since it is impossible to associate with this mechanism a state of stress in the soil which everywhere satisfies the condition that the maximum shearing stress is at most equal to c; this can be seen at once, since a failure surface must intersect a free boundary at an angle of $\frac{1}{4}\pi$ (in the case $\phi = 0$) if it is to coincide with the direction of maximum shearing stress at the boundary.

A lower bound, although rather a poor one, may be established from the simple stress field shown in fig. 5.17. In zone 1 it is assumed that $\sigma_x = 0$, $\sigma_y = \gamma y$ and $\tau = 0$, values which certainly satisfy the two equilibrium equations

$$\left.\begin{aligned} \frac{\partial \sigma_x}{\partial x} + \frac{\partial \tau}{\partial y} &= 0, \\[2mm] \frac{\partial \sigma_y}{\partial y} + \frac{\partial \tau}{\partial x} &= \gamma, \end{aligned}\right\} \tag{5.22}$$

and which also satisfy the loading boundary conditions on the free horizontal surface and the vertical cut. The maximum shearing stress in zone 1 is $\frac{1}{2}\gamma y$; if this is less than c, then y cannot exceed $2c/\gamma$.

The stresses marked in zone 2 also satisfy equations (5.22), and correspond to a maximum shearing stress of $\frac{1}{2}\gamma h$, while in zone 3 the shearing stress is zero. Thus for $h = 2c/\gamma$ the stresses in all three zones will satisfy both the equilibrium and the yield condition, and correspond therefore to a safe state: thus

$$h \geqslant 2c/\gamma. \tag{5.23}$$

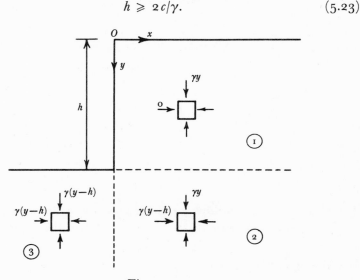

Fig. 5.17

It will be noted that certain *discontinuities* in stress occur across the boundaries dividing the three zones in fig. 5.17. Briefly, discontinuities are permitted only in the value of the direct stress parallel to such boundaries; the stress normal to the boundary and the shearing stress on planes parallel to the boundary must have the same values on each side. Such discontinuities do not affect the validity of the lower-bound theorem of plasticity.

DRUCKER (1953) (see also Drucker and Prager) has shown that if the medium is incapable of taking tension, so that the yield criterion of fig. 4.4 is modified to that shown in fig. 5.18, then the height h given by (5.23) is the actual limiting height. A trench dug in soil with these ideal characteristics would have sides which started to 'crack' when the depth reached the value $2c/\gamma$, fig. 5.19.

However, this is not the problem posed in this section, which is supposed to deal with a material capable of taking tension. PALMER has established recently a lower bound of $3c/\gamma$, but there is at present no solution bridging the fairly wide limits

$$3c/\gamma \leqslant h \leqslant 3{\cdot}83c/\gamma. \qquad (5.24)$$

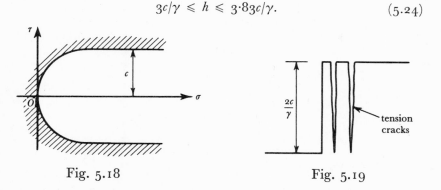

Fig. 5.18   Fig. 5.19

The stability of a vertical cut (2); (c, ϕ)

The results for a soil having both cohesion and internal friction may be developed from the work discussed in the last section. The upper bound given by the wedge analysis, fig. 5.20, is

$$h \leqslant \frac{4c}{\gamma[\sin 2\alpha - \tan\phi + \tan\phi\cos 2\alpha]}, \qquad (5.25)$$

corresponding to equation (5.19). A relative minimum occurs for α equal to Prony's angle $s\ (= \frac{1}{4}\pi - \frac{1}{2}\phi)$, and hence, corresponding to the bound (5.20),

$$h \leqslant 4\frac{c}{\gamma}\cot s. \qquad (5.26)$$

This bound may be improved by some 3 or 4 per cent by considering a curved failure surface. The appropriate curve is the logarithmic spiral, fig. 5.21, which automatically gives the correct dilatation at every point. Corresponding to equation (5.21), Chen, Giger and Fang give table 5.1 for various friction angles.

The lower bound corresponding to the stress pattern of fig. 5.17 is

$$h \geqslant 2\frac{c}{\gamma}\cot s, \qquad (5.27)$$

as may be deduced from equation (5.15); as before, the equality will hold in (5.27), that is, the solution will be correct, if the soil cracks in tension, fig. 5.19.

Table 5.1

Friction angle ϕ, degrees	0	5	10	15	20	25	30	35	40
N_s ($h = N_s c/\gamma$)	3.83	4.19	4.59	5.02	5.51	6.06	6.69	7.43	8.30
($4 \cot s$	4.00	4.37	4.77	5.21	5.71	6.28	6.93	7.68	8.58)

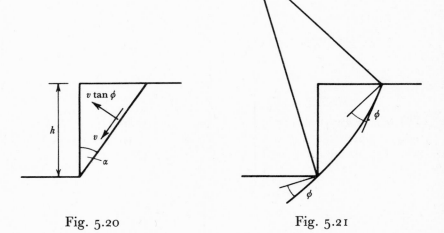

Fig. 5.20   Fig. 5.21

Thrust against a rough retaining wall; $c = 0$

It has already been noted that the value given by Coulomb for the thrust of a cohesionless soil against a smooth retaining wall is correct. By the upper-bound theorem of plasticity, the value A of the thrust cannot be less than $\frac{1}{2}\gamma h^2 \tan^2 s$, equation (5.7), since this value was derived by considering a possible mechanism of failure. Similarly, the Rankine analysis satisfies the lower-bound theorem, and leads to the same value of A, which is therefore the true value.

Coulomb modified his analysis on pp. 21–2; pp. 55–6 to allow for the effect of a rough retaining wall. He retained the plane-wedge mechanism, and obtained the general expression for the thrust (in the absence of cohesion):

$$A = \frac{1}{2}\gamma h^2 \frac{\tan\alpha(1 - \tan\phi\tan\alpha)}{m + \mu\tan\alpha}, \qquad (5.28)$$

144

where α is the wedge angle; if ϕ_1 is the angle of friction of the soil against the retaining wall, the constants m and μ are given by

$$
\left.
\begin{aligned}
m &= \tan\phi + \tan\phi_1, \\
\mu &= 1 - \tan\phi\tan\phi_1, \\
\frac{m}{\mu} &= \tan(\phi+\phi_1).
\end{aligned}
\right\}
\tag{5.29}
$$

The angle of the most critical wedge is given by

$$
\tan\alpha = -\frac{m}{\mu} + \left(\frac{m^2}{\mu^2} + \frac{m}{\mu\tan\phi}\right)^{\frac{1}{2}},
\tag{5.30}
$$

and, on substituting this value of $\tan\alpha$ back into equation (5.28),

$$
A_{\max} = \tfrac{1}{2}\gamma h^2 \{\sec\phi + [\tan\phi(\tan\phi+\tan\phi_1)]^{\frac{1}{2}}\}^{-2} \equiv Q.\tfrac{1}{2}\gamma h^2, \tag{5.31}
$$

which is probably the most compact form of the expression. Table 5.2 gives the values of Q defined by equation (5.31) for various values of ϕ and ϕ_1. The values $Q_{\text{Sok.}}$ in table 5.2 are those given by Sokolovskii, and are discussed below. Note that the value of $Q = \tfrac{1}{8}$ for $\phi = \phi_1 = 45°$ is that found by Coulomb in his numerical example, p. 22; p. 56.

Table 5.2

ϕ		10			20			30		
ϕ_1	0	0	5	10	0	10	20	0	15	30
Q	1.000	0.704	0.660	0.625	0.490	0.440	0.401	0.333	0.291	0.257
$Q_{\text{Sok.}}$	1.00		0.66	0.64		0.44	0.41		0.29	0.27

ϕ		40			45	
ϕ_1	0	20	40	0	$22\frac{1}{2}$	45
Q	0.217	0.187	0.161	0.172	0.148	$\tfrac{1}{8}$
$Q_{\text{Sok.}}$		0.19	0.17			

Equation (5.28) for the thrust may once again be established by writing a balance of virtual work for a displacement of the triangular wedge, care being taken to specify with precision the motion of the retaining wall; as will be seen immediately, the care required is excessive, and places an inadmissible restriction on the analysis. Figure 5.22 shows the wedge sliding along a plane inclined at an angle α to the vertical; the usual two velocity components are shown for the relative motion of the wedge and the undisturbed soil (cf. fig. 5.9). Thus the soil adjacent to the wall will have a vertical velocity

$$v(\cos\alpha - \tan\phi\sin\alpha)$$

and a horizontal velocity

$$v(\sin\alpha + \tan\phi\cos\alpha).$$

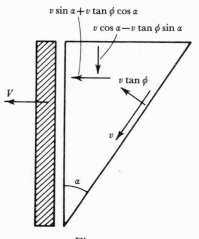

Fig. 5.22

Since the friction angle of the soil against the wall is ϕ_1, *there must be a dilatant velocity of the wall relative to the soil* of $\tan\phi_1$ times the sliding velocity. Thus, if the wall itself moves with a velocity V, fig. 5.22,

$$V - v(\sin\alpha + \tan\phi\cos\alpha) = \tan\phi_1 . v(\cos\alpha - \tan\phi\sin\alpha),$$

that is
$$V = v\cos\alpha(m + \mu\tan\alpha), \qquad (5.32)$$

where m and μ have the values (5.29). Thus the work balance gives

$$Av\cos\alpha(m + \mu\tan\alpha) = (\tfrac{1}{2}\gamma h^2\tan\alpha)\,[v(\cos\alpha - \tan\phi\sin\alpha)],$$

which leads at once to equation (5.28).

However, this plausible mathematical argument fails because the statement in italics at the head of the previous paragraph is not necessarily true. It is certain that for a material with *internal* friction, which obeys the Mohr–Coulomb criterion, a shearing deformation must be accompanied by dilatation, but there is no such compulsion governing the interaction of the soil with its external surroundings.

In particular, there is no mechanism for ensuring that the retaining wall will move with precisely the velocity V of equation (5.32). Thus, strictly, equation (5.28) and *a fortiori* equation (5.31) are both invalid. The fundamental requirements of plasticity theory are not satisfied in the presence of external friction, as was pointed out by DRUCKER; his theorems will be discussed in the next section of this chapter.

Even without this basic defect, Coulomb's analysis would still, of course, give only a lower bound on the value of the force on the retaining wall; it is not possible to associate a statically admissible stress field with the plane wedge of failure. For example, the Rankine field of fig. 5.10(a) demands zero shear stress on a vertical plane, whereas the presence of external friction implies a shearing stress between the soil and the wall. However, this point is trivial numerically; the values $Q_{Sok.}$ in table 5.2 are 'exact' (but strictly meaningless from the point of view of plasticity theory), and are close to Coulomb's own values in the table.

An exception arises if the external friction angle ϕ_1 is greater than the internal friction angle ϕ, and if the soil is imagined to be 'glued' to the wall so that friction can in fact be mobilized (in the *soil*) as a failure mechanism develops. Equation (5.30) then gives

$$A_{max} = \tfrac{1}{2}\gamma h^2(\sec\phi + \sqrt{2}\tan\phi)^{-2}.$$

Drucker's theorems

DRUCKER's 'anthropomorphic' presentation (1954) of the theorems of plasticity is:

The two major limit theorems state that an assemblage of elastic – perfectly plastic bodies, with zero friction or complete attachment at each interface, will on the one hand, do the best they can to distribute stress to avoid collapse. On the other hand, they will recognize defeat if any kinematic collapse mode exists.

He then demonstrates that if friction is finite and non-zero on at least one of the interfaces then the system is no longer 'intelligent enough to distribute the stress to avoid collapse'.

This failure of the theorems unfortunately leads to unsafe calculations. Thus if the effect of external friction is apparently to reduce the value of the load acting on a retaining wall, such a reduction cannot be relied on; the external source of friction may, to continue the anthropomorphic description, get out of the way and refuse to lend its support to the collapsing assemblage. However, Drucker establishes three theorems which are of some help in interpreting cases of failure which involve external friction.

A. Any set of loads which produces collapse for the condition of no relative motion (i.e. no sliding or separation) at the interfaces will produce collapse for the case of finite friction.

B. Any set of loads which will not cause collapse when all coefficients of friction are zero will not produce collapse with any values of the coefficients.

C. Any set of loads which will not cause collapse of an assemblage of bodies with frictional interfaces, will not produce collapse when the interfaces are 'cemented' together with a cohesionless soil of friction angle ϕ_1.

Thus in table 5.2, the values for $\phi_1 = 0$ can at least be relied on to give *relatively* safe values of the coefficient Q for any actual value of ϕ_1; for $\phi = 45°$, for example, although the value $Q = \frac{1}{8}$ for $\phi_1 = 45°$ is suspect, at least it cannot exceed 0.172 for the same class of mechanism. In fact, since the solutions in table 5.2 for $\phi_1 = 0$ happen to be correct (that is, the mechanism of failure does indeed involve a plane surface), the value $Q = 0.172$ is *absolutely* safe.

Nineteenth-century work

PRONY's *Recherches sur la poussée des terres* of 1802 has already been noted, as has his use of angles rather than lengths as unknown quantities. This leads to great simplification of the working, but Prony's *theory* is exactly that of Coulomb. His concern is to make simple and straightforward calculations, and his short book (44 pages) concludes with a graphical construction for the dimensions of retaining walls.

MAYNIEL's *Traité expérimental* is invaluable as a summary of all the major work before 1808. The treatise is divided into four books, of which the first presents and discusses the results of previous tests on soil thrust, including those of Gadroy, Papacino, Gauthey and Rondelet. Further, Mayniel reports tests made by himself and other military engineers in 1805 at Alessandria (Piemonte) and in 1806 and 1807 at Jülich (Rheinland), all directed towards measuring the thrust on a retaining wall. The scale was quite large, with a box about 3 m long and cross-section 2 m × 2 m; as usual, the end of the box was hinged, and the results of 33 tests are given.

In book 2 Mayniel reviews historically the various theories of earth pressure, especially that of Coulomb. He reworks Coulomb's analysis in terms of angles rather than lengths, and considers the 'general case' when neither face of the retaining wall is vertical. He concludes, in book 3, by comparing the generalized Coulomb theory and the tests at Jülich, that Coulomb's theory is the best; moreover, it is the simplest. There follows a long list of different cases, including buttressed retaining walls, and book 4 gives rules and examples for practical construction.

Neither Prony nor Mayniel is really critical of Coulomb's work; they both allow for cohesion as well as internal friction in their analyses, but it seems clear that they are really thinking of cohesion-less behaviour. This attitude received its fullest expression in the work of PONCELET (1839), who assumed a purely frictional soil. He devised an ingenious graphical construction for the determination of Coulomb's plane of failure for a fill with a sloping free surface acting on an inclined retaining wall. Such theories of cohesionless behaviour proceeded sometimes successively and sometimes independently, and advances were made by RANKINE (1857) and by LÉVY (1873), and by CONSIDÈRE (1870) and BOUSSINESQ (1876 and later).

RANKINE's work has already been noted; he established the planes of rupture as coinciding with the characteristics, and refers to Coulomb's determination of these same planes by the use of the 'ideal wedge of least resistance'. LÉVY's *Mémoire* was read in 1867 to the *Académie*, and published in its final form in 1873 in Liouville's *Journal*. It was written without knowledge of Rankine's paper of ten years earlier, and Lévy covers, with great clarity, much the same ground as Rankine. He establishes the relation between principal stress for a cohesionless soil in limiting equilibrium,

$$\sigma^2 = \sigma_1 \tan^2 s \tag{5.33}$$

(cf. equation (5.10) and fig. 5.12), and that the slip curves (*courbes de glissement*) always make a constant angle ($\frac{1}{4}\pi \pm \frac{1}{2}\phi$) with the principal directions (*courbes isostatiques*). He deduces that for problems in which the principal directions are parallel straight lines, then the slip lines (*lignes de glissement*) must also be parallel straight lines.

Further, Lévy gives a full discussion (which Rankine did not) of the case where there is friction between the soil and the wall. He notes that the ratio of shear stress to normal stress at the face of the wall is established *a priori* at failure, being equal to $\tan\phi_1$, where ϕ_1 is the friction angle ($< \phi$) between the soil and masonry; except by accident, this ratio will be incompatible with the values arising from the Mohr–Coulomb analysis of the soil itself. If the back face of the wall is inclined at a particular angle, however, which Lévy calculates in terms of the slope of the free surface of the soil (not necessarily horizontal) and of the angles ϕ and ϕ_1, then the Coulomb analysis is valid; otherwise the principal directions will no longer remain straight, and the current state of the theory does not permit him to make an exact analysis.

He proposes, therefore, that the back of the retaining wall should be given a batter (*fruit*) of precisely the correct amount to make the theory with straight slip lines valid; this inverse method of design should have commended itself to Saint-Venant! SAINT-VENANT in fact reported to the *Académie* on Lévy's paper, and his comments are reprinted in Liouville's *Journal* for 1870; the report is followed by a short paper by Saint-Venant himself. Lévy had already noted that a potential function ψ could be introduced to satisfy the equilibrium equations (5.22):

$$\sigma_x = \frac{\partial^2 \psi}{\partial y^2}, \quad \sigma_y = \gamma y + \frac{\partial^2 \psi}{\partial x^2}, \quad \tau = -\frac{\partial^2 \psi}{\partial x\, \partial y}. \tag{5.34}$$

If these values are substituted into the yield criterion

$$4\tau^2 + (\sigma_y - \sigma_x)^2 = \sin^2\phi(\sigma_x + \sigma_y)^2, \tag{5.35}$$

an equation for ψ results. Saint-Venant comments that the equation is of second order and second degree, and one can never hope for a general exact solution.

However, an exact solution may indeed be found, for part at least of the field, for the case where the free surface of the soil is plane, whether sloping or not, and this was in fact the solution determined by Lévy. In fig. 5.23 the free surface makes an angle ω with the horizontal, and the back face of the wall an angle ϵ_1 with the vertical. Saint-Venant shows that equations (5.34) and (5.35) are satisfied by

$$\left.\begin{aligned}
\sigma_x &= \gamma p \lambda^2 \cos\omega, \\
\sigma_y &= \gamma p \left(\frac{1 + \lambda^2 \sin^2\omega}{\cos\omega}\right), \\
\tau &= -\gamma p \lambda^2 \sin\omega,
\end{aligned}\right\} \tag{5.36}$$

where the parameter p (marked in fig. 5.23) is given by

$$p = y\cos\omega + x\sin\omega, \tag{5.37}$$

and where
$$\lambda = \frac{\cos\omega}{\cos\phi} - \left(\frac{\cos^2\omega}{\cos^2\phi} - 1\right)^{\frac{1}{2}} \tag{5.38}$$

(note that for $\omega = 0$, $\lambda = \sec\phi - \tan\phi = \tan s$, and $p = y$). As Lévy had said, equations (5.36) will give the exact solution providing the angle ϵ_1 has a certain definite value (for which both Lévy and Saint-Venant give complicated expressions) such that the ratio τ/σ at the face of the wall is exactly $\tan\phi_1$. For the particular case $\phi_1 = \phi$, the value of ϵ_1 may be found from

$$\cos\left(2\epsilon_1 + \phi - \omega\right) = \frac{\sin\omega}{\sin\phi};$$ (5.39)

for $\omega = 0$, for example, $\epsilon_1 = \frac{1}{4}\pi - \frac{1}{2}\phi = s$, and for $\omega = \phi$, $\epsilon_1 = 0$.

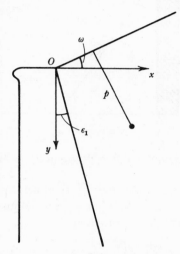

Fig. 5.23. After Saint-Venant (1870).

For all other values of ϵ_1, for which equations (5.36) are not valid over the whole field, Saint-Venant proposes nevertheless that the values should be used to give the design of the wall: 'All it comes to is to suppose that the angle of friction at the face of the wall, instead of being ϕ, *is always the smaller angle ϕ_1*...' (given by the ratio of shear stress to normal stress from equations (5.36); Saint-Venant's italics). Saint-Venant states that this procedure will always be safe; thus he asserts (but does not prove) Drucker's theorem C.

CONSIDÈRE published in 1870 a note on soil thrust which was originally written in ignorance of Lévy's work; in his introduction Considère gives complete priority to Lévy, and makes no claim that his own paper makes any further advance. He too establishes that slip lines will only be straight under a unique inclination of the back of the retaining wall, but he has no method for calculating the general case.

BOUSSINESQ published several papers on the thrust of cohesionless soil, and it is convenient to discuss first the *Mémoire* of 1882. Here the more general results of his 1876 paper (see below) are applied to a single problem, that of Coulomb, i.e. a mass of soil with a horizontal

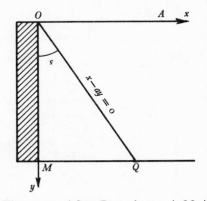

Fig. 5.24. After Boussinesq (1882).

free surface supported by a rough vertical retaining wall. In fig. 5.24 the angle of friction along OM is ϕ_1, and Boussinesq states clearly the equations governing the problem:

$$\left.\begin{aligned}
\frac{\partial \sigma_x}{\partial x} + \frac{\partial \tau}{\partial y} &= 0, \\
\frac{\partial \sigma_y}{\partial y} + \frac{\partial \tau}{\partial x} &= \gamma, \\
\sin^2\phi &= \frac{(\sigma_x - \sigma_y)^2 + 4\tau^2}{(\sigma_x + \sigma_y)^2},
\end{aligned}\right\} \tag{5.40}$$

together with the boundary conditions

$$\left.\begin{aligned}
\tau &= 0 \quad \text{and} \quad \sigma_y = 0 \quad \text{for} \quad y = 0 \\
\text{and} \qquad \frac{\tau}{\sigma} &= \tan\phi_1 \quad \text{for} \quad x = 0.
\end{aligned}\right\} \tag{5.41}$$

He proposes a solution to these equations in two parts, divided by the line OQ ($x = ay$, where $a = \tan s$) in fig. 5.24. For region OAQ he applies Rankine's solution

$$\sigma_x = \gamma y \tan^2 s, \quad \sigma_y = \gamma y, \quad \tau = 0, \tag{5.42}$$

which is valid for $x > y\tan s$, and for region OMQ he proposes the following linear functions of x and y for the values of the stresses:

$$
\left.
\begin{aligned}
\sigma_x &= \frac{\gamma\tan^2 s(y + x\tan\phi_1)}{1 + a\tan\phi_1}, \\[2mm]
\sigma_y &= \frac{\gamma(y + x\tan\phi_1)}{1 + a\tan\phi_1}, \\[2mm]
\tau &= -\frac{\gamma a\tan\phi_1(ay - x)}{1 + a\tan\phi_1}.
\end{aligned}
\right\}
\tag{5.43}
$$

Equations (5.42) satisfy all three of (5.40) and the first of (5.41); equations (5.43) satisfy the first two of (5.40) and the second of (5.41). Thus equilibrium is satisfied throughout the field, as are the boundary conditions on OA and OM; there is a weak discontinuity along OQ, the values of the stresses for $x = ay$ being in fact the same from (5.42) and (5.43).

The yield condition is satisfied for the Rankine region, but not for the region OMQ; instead, substitution of the values (5.43) into the third of (5.40) gives a fictitious angle of internal friction

$$
\sin^2\phi' = \sin^2\phi + (1 - \sin\phi)^2\tan^2\phi_1\left(\frac{ay - x}{ay + ax\tan\phi_1}\right)^2. \tag{5.44}
$$

For smallish values of ϕ_1 the fictitious angle ϕ' will differ very little from (but will always exceed) the real value ϕ in the soil, and Boussinesq discusses at length, with practical calculations, the reliance that can be placed on the values of forces obtained from this analysis.

The importance of this analysis lies, of course, in the introduction of a discontinuity within the stress field, and Boussinesq examines this problem more fully and more generally in his earlier paper of 1876. In fig. 5.25 the (plane) free surface OA of the soil makes an angle ω with the horizontal, and the back of the retaining wall OM an angle ϵ_1 with the vertical (cf. fig. 5.23). If an auxiliary angle ω is introduced, being the solution between 0 and s (for positive ω) of the equation

$$
\sin(\omega + 2\psi) = \frac{\sin\omega}{\sin\phi}, \tag{5.45}
$$

then Boussinesq shows that Lévy's proposal for the batter of OM is

$$
\epsilon_1 = s - \psi \tag{5.46}
$$

(cf. equation (5.39)). More generally, then, the line of discontinuity OQ in the stress field will lie as shown in fig. 5.25, and Boussinesq obtains approximate solutions for the values of the stresses in the region OMQ, analagous to those of (5.43), which are continuous with the 'Rankine' stresses in OAQ across OQ, but whose derivatives are discontinuous.

Boussinesq states clearly that he is assuming that the *whole* of the soil within the field considered is in the limiting state, and he remarks particularly, both in the 1876 and in the 1882 paper, on the case

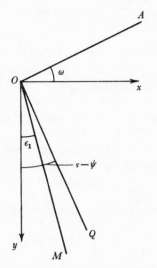

Fig. 5.25. After Boussinesq (1876).

when the angle ϵ_1 in fig. 5.25 exceeds $(s - \psi)$, so that OM lies to the right of OQ. It is then impossible to satisfy the second boundary condition of (5.41), so that,

when a rough retaining wall, having its back face inclined at an angle greater than $(s - \psi)$, *starts* to overturn, the state of slip (*l'état ébouleux*) is not produced throughout the medium. No doubt *a wedge of soil, with its base up, remains attached to the wall and separates as a whole from the rest of the medium, following a rupture line parallel to OQ.* [Boussinesq's italics.]

Boussinesq obtains as a result of his theory a general expression for the force tending to overturn the retaining wall. For the particular case of a vertical wall and a horizontal fill ($\omega = \epsilon_1 = 0$), and for $\phi_1 = \phi$, the overturning couple is

$$M = \tfrac{1}{6}\gamma h^3 \tan^2 s \left[\frac{\cos s \cos \phi}{\cos(\phi - s)} \right], \tag{5.47}$$

where the term in square brackets modifies the Coulomb–Rankine value, equation (5.9). DARWIN reported in 1883 seven series of tests on sand, in which he quotes the results of equations (5.9) and (5.47), and attempts to distinguish between them experimentally. He found in all cases that Rankine's theory, which ignores friction against the wall, gave large discrepancies, but that Boussinesq's formula gave results much closer to those observed. However, Darwin found that the 'historical element' (suggested to him by Clerk Maxwell) led to some scatter; that is, the values of thrust were dependent on the way sand was packed into the box, and he concluded that Boussinesq's theory was not altogether satisfactory.

Darwin's experiments were also reported in the *Annales des Ponts et Chaussées*, and led to considerable discussion; in particular, Boussinesq made a contribution in which he found that the experiments 'appear to confirm as fully as possible' his own formulae. The difficulty in interpreting the experimental results lay in the correct choice of the angle ϕ of internal friction. Boussinesq argues plausibly that the value will vary through the mass of sand, and that only a small change in average value will lead to almost exact concordance between theory and experiment. However, Boussinesq had solved his equations (5.40) and (5.41) on the assumption of a constant value of ϕ, and the use of an average value to approximate a quantity which may be varying continuously, while perhaps leading to reasonably accurate practical calculations, does not seem to be satisfactory scientifically; he appears not to have pursued the idea of a *theory* based on a variable value of ϕ. A paper in 1884 returns to an examination of his basic equation (5.47) for a vertical wall with a horizontal fill, but really adds little to his previous work.

According to KERISEL, it was not until 1934 that CAQUOT gave complete solutions to Boussinesq's equations (5.40), that is, to Lévy and Saint-Venant's problem of equations (5.34) and (5.35). More recently, SOKOLOVSKII has obtained numerical solutions for a large number of problems, and some of these are discussed below. Sokolovskii may be regarded as the final exponent of that branch of soil mechanics, which, having already grossly simplified the problem to one involving only two parameters, c and ϕ, then proceeds to investigate the consequences of purely frictional behaviour. This cohesionless theory has been attacked repeatedly as a poor model of the behaviour of real soil; ALEXANDRE COLLIN, for example, published a book in 1846 on landslides in clays, in which he makes it clear that shear strength is one of the fundamental parameters that must be used to understand the failure of clay slopes (see, for example,

SKEMPTON, 1950). He saw that for such failures the rupture surface would be curved, and modern slip-circle analysis (such as that of PETTERSON and of FELLENIUS, see for example SCHOFIELD and WROTH) may be traced back to Collin. Similarly, BENJAMIN BAKER criticized the cohesionless theory in 1881; indeed, for a large number of problems a $\phi = 0$ (i.e. frictionless) analysis is more appropriate than a $c = 0$ analysis. The $\phi = 0$ analysis is, of course, a branch of the pure theory of plasticity and, as such, will not be discussed further here.

Sokolovskii

In fig. 5.26 one principal direction at a particular point in the soil under a given state of stress is shown making an angle ψ with the x-axis. The soil is assumed to be in its limiting state, and to possess both cohesion and internal friction, so that not only must the first two of equations (5.40), the equilibrium equations, be satisfied, but also the general yield criterion

$$\sin^2\phi = \frac{\frac{1}{4}(\sigma_x - \sigma_y)^2 + \tau^2}{[\frac{1}{2}(\sigma_x + \sigma_y) + c\cot\phi]^2} \tag{5.48}$$

(cf. the third of equations (5.40)). Two variables, σ_0 (fig. 5.27) and χ, are introduced, where

$$\sigma_0 = \frac{1}{2}(\sigma_x + \sigma_y) + c\cot\phi, \tag{5.49}$$

and
$$\chi = \frac{1}{2}\cot\phi\log\sigma_0. \tag{5.50}$$

The equilibrium equations for the soil in the limiting state may now be written referred to new axes (α, β) making angles s ($= \frac{1}{4}\pi - \frac{1}{2}\phi$) with the principal direction, fig. 5.26:

$$\left.\begin{array}{l}\dfrac{\partial}{\partial\alpha}(\chi + \psi) = \frac{1}{2}\gamma\exp(-2\chi\tan\phi)\cos(\psi - s)/\sin\phi, \\[2mm] \dfrac{\partial}{\partial\beta}(\chi - \psi) = -\frac{1}{2}\gamma\exp(-2\chi\tan\phi)\cos(\psi + s)/\sin\phi.\end{array}\right\} \tag{5.51}$$

(This transformation may be found in Sokolovskii, or perhaps may be followed more easily in DE JOSSELIN DE JONG.)

Equations (5.51) form the basis of a method of numerical integration along the characteristic lines α and β. Suppose that the state of stress is known at point A in fig. 5.28(a); then the value χ_A is calculable immediately from (5.49) and (5.50), the principal direction ψ_A will be known, and hence the α-direction at A (fig. 5.26). The first of equations (5.51) will then give, approximately, the value of $(\chi_C + \psi_C)$ at a neighbouring point C. Similarly, values at B may be introduced into the second of (5.51) to give the variation of $(\chi - \psi)$ along the β-direction, so that $(\chi_C - \psi_C)$ may be calculated. Thus numerical values of χ and ψ may be found within a certain region of the field; if PQ in fig. 5.28(b) represents (for example) a boundary of the soil along which values of the stresses are known, then the stresses may be evaluated within the region PQR, but not outside this region.

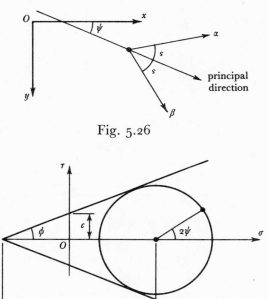

Fig. 5.26

Fig. 5.27

Some of the features of Boussinesq's analysis of soil thrust may now be discussed. Starting from the free surface OA of the soil, fig. 5.29 (cf. fig. 5.24), it is easily shown that the characteristics in the triangular region OAQ must be straight. If the back face of the wall is frictionless, then the characteristics must also be straight within the region OQM, and the network may be completed as in fig. 5.30. This is the Coulomb–Rankine solution, *whether the material has cohesion or not*; as has been seen, the solution is actually correct only for the case $c = 0$.

If the back face of the wall is rough, and if it is assumed that the consequent frictional forces can in fact be mobilized, then the characteristics must intersect OM in fig. 5.30 at certain definite angles in order that the appropriate ratio τ/σ should be equal to $\tan\phi_1$. The characteristics in the region OQM will then, in general, be curved, and there will be a discontinuity along OQ, exactly as postulated by Boussinesq. Figure 5.31 shows a typical result obtained by Sokolovskii, in which there are *three* regions; a *fan OQR* separates the two regions

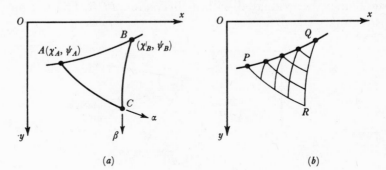

(a)   (b)

Fig. 5.28

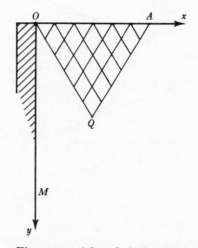

Fig. 5.29. After Sokolovskii.

OAQ and ORM. In such a fan, one set of characteristics is straight, and the other forms the familiar logarithmic spirals.

Figure 5.32 shows (again from Sokolovskii) the results of a $c = 0$ analysis for three different friction angles at the wall, all for $\phi = 30°$; the numerical values of thrust ($Q_{\text{Sok.}}$) resulting from this and similar analyses have been entered in table 5.2 above. Immediately following this particular piece of work. Sokolovskii states that he will generalize his results for a medium which possesses cohesion as well as internal friction. However, he only works cases in which the surface OA is subjected to a (uniform) overburden; fig. 5.31, for example, is drawn for the case $p = 5c$, where p is the uniform pressure resulting from the overburden.

Sokolovskii concerns himself, in fact, with problems in which the *whole* of the region (such as $OAQRM$ in fig. 5.31) is in a critical state.

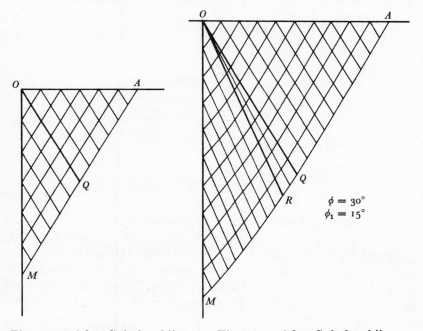

Fig. 5.30. After Sokolovskii.   Fig. 5.31. After Sokolovskii.

Without an overburden, it is not possible to bring the region neighbouring O into the yielding condition, and part at least of the soil would slip against the overturning retaining wall as a rigid body, as foreseen by Boussinesq. Similarly, in his analysis of the stability of slopes (whether vertical cuts or not) Sokolovskii again uses the device of an overburden so that equations (5.51) may be applied throughout the region considered. Sokolovskii is not alone in neglecting such 'mixed' problems involving both finite critical zones and rigid regions; it is this neglect which has led to the failure to narrow the poor bounds (5.24) on the greatest depth of a vertical cut.

A second and fundamental criticism of Sokolovskii's work is that his solutions are incomplete. In fig. 5.31, for example, the solution is obtained only in the region $OAQRM$, and is not extended to the rest of the medium, which is *assumed* to remain unyielding. Sokolovskii effectively postulates a mechanism of failure, and chooses to evaluate the quantities involved· by statics (rather than by a work equation, for example), so that, within the failure region, both the equilibrium and the yield conditions are also satisfied. However, this satisfaction of yield and equilibrium does not strictly make Sokolovskii's solutions

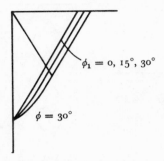

$\phi_1 = 0, 15°, 30°$

$\phi = 30°$

Fig. 5.32. After Sokolovskii.

safe. Even if the problems involve no external friction, and hence are covered by the plasticity theorems, a full solution should show that yield and equilibrium are satisfied everywhere. However, this is not really a practical criticism. At reasonable depths the self-weight of the soil will lead to high frictional forces, which will effectively prevent the formation· of mechanisms remote from the region considered.

Critical state theory

It has been the intention of this chapter to trace the development of a particularly simple idea, and to show the application of the Mohr–Coulomb criterion, fig. 5.13, to the problems Coulomb himself considered. As a piece of applied mathematics the work is, as has been seen, far from complete; a much better (c, ϕ) analysis is required for the stability of a vertical cut (let alone a sloping surface) and Coulomb's complete problem of the thrust of a (c, ϕ) soil against a rough retaining wall will also reward fuller investigation.

These problems are well worth pursuing in their own rights, but their solutions are not likely to be of the first importance to the development of soil mechanics. The Mohr–Coulomb criterion is really far too simplistic to form a proper model for the solution of more general problems. Two parameters, c and ϕ (even if they are allowed to vary), are not enough to describe adequately the behaviour of soil. A modern attempt at a more sophisticated formulation is that of critical state theory, due to ROSCOE, SCHOFIELD and WROTH (see also SCHOFIELD and WROTH). Here *three* parameters are used to specify the soil properties, and the yield surface cannot be described in two dimensions.

The third parameter is concerned with the water content of the soil, and a general soil must be treated as a two-phase material. BOSSUT was well aware of pore-water pressures at the base of a dyke, and Coulomb, as a practical engineer, was aware that soil properties are altered markedly by the presence of water (p. 23; p. 57). When the presence of water is taken into account, some of the anomalies of the (c, ϕ) theory disappear. Critical state theory answers, for example, a serious practical objection raised to the application of the Mohr–Coulomb theory to clays. As has been seen in this and the previous chapter, a shear failure in a (c, ϕ) soil must be accompanied by large dilatations (as in fig. 5.9). Such large volume changes are not observed in clays (although they do occur in sand), and this is readily explained by critical state theory.

More recently, CALLADINE has made a conceptual model of soil which involves ideas stemming directly from the theory of plasticity, but which does not introduce water content as such. Behaviour predicted by Calladine's model is close, however, to that of real soils, so that there is hope of embracing some at least of the problems of soil mechanics within a discipline which is now well established.

6

The Thrust of Arches

'La poussée des voûtes' and 'la poussée des terres' are two topics
that recur continually in the eighteenth century. It was seen in the
last chapter that values of soil thrust were required in order to compute the dimensions of retaining walls; this was Gautier's fifth
problem. His first problem concerned the thickness of abutment piers,
and it was for this purpose that values of arch thrust were needed;
the study of arches was directed initially to the determination of the
sizes of the abutments, rather than to the action of the voussoir arch
itself. However, it was soon appreciated that some assumptions had
to be made about the behaviour of the voussoirs in order to obtain
an estimate of the abutment thrust.

COUPLET, whose two *Mémoires* on arches (1729 and 1730) are discussed more fully below, starts by referring to his own work on soil
thrusts, and he notes the similarity of the two studies.

Quoique cette matiére paroisse différente de l'autre, elle n'est cependant, à le
bien prendre, que la même, ou du moins l'une paroît être une suite nécessaire de
l'autre; c'est pourquoi je crois placer ces nouvelles recherches en leur lieu, en les
mettant à la suite de la poussée des Terres.

Ceux qui ont proposé ces deux Problemes d'Architecture, en ont bien senti la
liaison, en les proposant tous deux ensemble; & ceux qui ont entrepris leur solution, ont confirmé l'identité de cette matiére, en travaillant a la solution de l'un
& de l'autre.

Thus, according to Couplet, not only are the problems analogous,
but so also are the solutions. In his calculation of soil thrusts, Couplet
did not take into account *internal* friction, and he specifically states
that he considered soils 'comme roulantes & détachées les unes des
autres'; it was precisely the opposite assumption which led, as will
be seen, to his major contribution to the problem of the thrust of
arches.

The plastic theorems applied to masonry

The key to the understanding of the behaviour of masonry lies in
Coulomb's *Remark I* (p. 38; p. 68) that friction is often large enough
in the materials used for arch construction that the different voussoirs
could not slide one on another. If this be accepted, then Coulomb's

proposed mode of failure at a cross-section, that of *hinging* about a free edge, seems to be the only possible mode. Assuming further, with Coulomb (p. 38; p. 68), that cohesion of the joints can be neglected, then the horizontal thrust necessary for the stability of an arch is found to lie between two limits; in Coulomb's fig. 14, these limits correspond to rotation about a point M on the intrados or a point m on the extrados, and have values $\phi . gM/MQ$ and $\phi . qg'/mq$ respectively.

Now these expressions for the value of the arch thrust involve the density of the material and the dimensions of the arch, but contain no terms corresponding to the strength of the material; they are thus purely geometrical statements of the stability of the arch. It is true that the actual fracture joint Mm has to be determined, but again this is a problem to be solved in geometrical terms (as Coulomb notes in his *Remark II* (pp. 38ff; p. 69), trial and error is a quick and easy method). Thus the assumption of a hinging mode of failure has led to the conclusion that material strength is not of importance in the analysis of arches; there is no question, for example, of crushing of the stone. The generally low level of stress in masonry construction will be discussed a little later; for the purpose of establishing the basic principles of the action of masonry, the following formal assumptions will be made (Heyman 1966, 1969):

(i) *Stone has no tensile strength.* Although stone itself may have some tensile strength, the joints between voussoirs may be dry or made with weak mortar. Thus the assumption is equivalent to the statement that no tensile forces can be transmitted within a mass of masonry. In accordance both with common sense and with the principles of the plastic limit theorems (which are discussed below), this assumption is 'safe'; it may be too safe, that is, unrealistic, if, for example, spandrel masonry interlocks with the voussoirs in an arch so that tensile forces can be transmitted locally by such stones.

(ii) *Stone has an infinite compressive strength.* This is equivalent to the assumption that stresses are so low in masonry that there is no danger of crushing of the material. The assumption is obviously unsafe, but the errors introduced are very small in the analysis of any real structure of the type considered here.

(iii) *Sliding failure cannot occur.* It will be assumed that friction is high enough, or that the stones are effectively interlocked, so that they cannot slide one on another. (Heyman (1968) has given an example (of buttressing for a large cathedral) where sliding failure may be of importance.)

With these assumptions KOOHARIAN, acting on a suggestion of DRUCKER (see also PRAGER, 1959), has shown that masonry may be treated as a material to which the limit theorems, developed in the last few years for the analysis of the plastic behaviour of ductile steel frames (see, for example, BAKER, HORNE and HEYMAN), may be applied. Suppose a cross-section of a voussoir arch is at the point of failure by hinging about one edge, fig. 6.1(*a*). If the axial load carried by the arch has value N at the hinge position, then the hinge

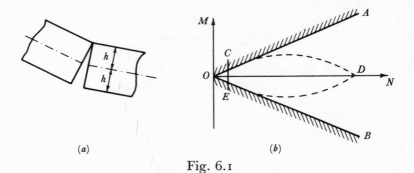

(*a*)   (*b*)

Fig. 6.1

will form at an effective bending moment $M = hN$. The lines $M = \pm hN$ define a permissible region, fig. 6.1(*b*), which is reminiscent of the Mohr–Coulomb criterion; however, fig. 6.1(*b*) is much simpler than say fig. 5.12. This is because M and N are *stress resultants* rather than stresses; there is no question here of a Mohr's circle being involved.

Instead, any point within the open triangle AOB in fig. 6.1(*b*) represents a state of a particular cross-section which is safe. That is, a safe state is one for which the thrust line lies within a distance h of the centre line of the masonry; under these conditions, no hinge will form, and no movement will occur at the joint. A point lying on the line OA or OB represents the formation of a hinge, and the thrust line lies in the surface of the masonry. A point outside the triangle AOB corresponds to the impossible state of the thrust line lying outside the masonry.

If the stone of infinite strength is replaced by the real stone with a finite crushing strength, then the yield surface AOB is replaced by the curved boundary $OCDEO$, formed by two parabolic arcs. This new yield surface has the same characteristic properties as those of the open triangle AOB; a point lying within the surface represents a safe state, and so on. However, only a small portion of this complete yield surface applies to any practical masonry structure. A typical value of permitted stress used in the nineteenth-century design of large bridges is 10 per cent of the crushing strength (YVON VILLARCEAU); in fact, nominal stresses will be likely to be even less than this, but even at 10 per cent the portion of the yield surface is confined to the slightly curvilinear triangle OCE. The sides OC and OE are so nearly straight in this region that little error is introduced by taking them to coincide with OA and OB. Thus the line of thrust cannot in fact approach a free edge to within closer than 5 per cent of the depth of the section, if the nominal stress is 10 per cent of the crushing strength; in all further discussion here, it will be assumed that hinging actually occurs about the edge of the masonry.

These assumptions and approximations allow the limit theorems of plastic design, which have already been used in chapters 4 and 5 above, to be applied to masonry. Of prime importance is the *safe theorem*. It has been seen that the line of thrust is confined at every section to lie within the depth of the masonry; it can be proved that if a thrust line can be found, for a complete arch, which is in equilibrium with the external loads (including the self weight of the arch), and which does indeed lie within the masonry, then the arch is safe. The power of this theorem lies in the fact that the thrust line found to satisfy the theorem need not be the actual thrust line.

Thus to demonstrate the stability of the arch it is necessary to show only that there is at least one internal force system for which it is actually stable; this gives complete assurance that the arch cannot collapse by any mode under the particular system of loads investigated. This is precisely the conclusion reached by Coulomb; in his discussion (pp. 36ff; pp. 67ff) of the limits A_1 and A' etc. between which the horizontal thrust must lie if equilibrium is to be attained, he makes no attempt to calculate the actual thrust. (It was noted that Gregory in 1697 effectively invoked the safe theorem when he stated that an arch of a shape other than a catenary can stand only if a 'catenary' can be included within its thickness.)

The formation of a single hinge does not, of course, necessarily make a masonry structure unstable. If a voussoir arch were fitted initially exactly between its abutments, but those abutments subsequently spread, then the arch would accommodate itself to the increased span by forming three hinges, one at the crown in the extrados, and two in the intrados, probably at the abutments if the arch is relatively flat (fig. 6.2(*a*)). At a hinge, the line of action of the thrust at the section is known; it must pass through the hinge point. Thus if three hinges are formed, then three points will be known on the thrust line for the arch, and this thrust line can then be constructed uniquely, as sketched in fig. 6.2(*b*). (It remains to be shown that the thrust line, compelled to pass through the three hinge points, can in fact be completed so that it lies within the masonry at every cross-section; however, this is always possible for *some* arrangement of three hinges, and it is assumed that the correct hinge pattern has been found.)

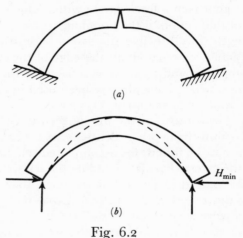

(*a*)

(*b*)

Fig. 6.2

The arch thrust, H_{min}, corresponding to the spread of the abutments, is the minimum that will ensure stability of the arch as a whole. Similarly, if the abutments are too close, fig. 6.3, three hinges will again form, but of opposite signs to the previous hinges; the corresponding thrust line gives an arch thrust, H_{max}, which is the largest that the arch can support. It is clear from figs. 6.2 and 6.3 that the values of H_{max} and H_{min} corresponding to the thrust lines shown can be very different, and that there may therefore be a wide range of possible stable configurations; further, very small movements of the abutments of an arch can lead to a large change in the value of the abutment thrust. Since the conditions at the springings of an actual arch will not be known precisely (or, if known at one particular time, they will be certain to change with the passage of years), it is to some extent meaningless to enquire as to the *actual* value of

166

the abutment thrust. Thus the safe theorem is not such an imprecise tool as might at first be supposed.

In figs. 6.2 and 6.3 the formation of three hinges led to limiting locations for the position of the thrust line. In terms of conventional notions of statical indeterminacy, a structure having three redundancies has been made statically determinate by the insertion of three hinges. However, such a determinate structure is not, of course, on the point of failure; for collapse to occur, a fourth hinge must be

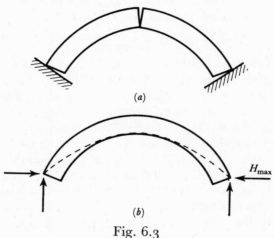

(a)

(b)

Fig. 6.3

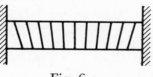

Fig. 6.4

formed to transform the arch into a *mechanism* (in this case, a four-bar chain). The geometry of the arch may be such that there is no possible location for such a fourth hinge (in which case, within the assumption of a stone of infinite crushing strength, the arch is infinitely strong); for example, a very flat arch (in the limit, the plate-bande of fig. 6.4) will allow of no arrangement of four hinges in the upper and lower surfaces which will give a mechanism.

Thus a duality becomes apparent from these considerations of stability. To demonstrate that an arch is stable, the safe theorem requires that a thrust line, in equilibrium with the external loads, can be constructed to lie wholly within the masonry. To demonstrate that an arch can become unstable, it is necessary to construct a pattern of hinges that corresponds to a mechanism of collapse. There is a formal uniqueness theorem, of some interest when considering minimum dimensions of arches, which considers a satisfactory thrust line which at the same time admits sufficient hinges to turn the arch into a mechanism. All these ideas, which are no more than the 'translation' of theorems of plasticity in terms of masonry, can be used to discuss the proper design of arches, and serve to illuminate the findings of Coulomb and his predecessors.

Eighteenth-century work (1)

HOOKE was evidently aware in 1675, with his *ut pendet...* (see p. 76 above), that the solution of the problem of the hanging chain would also give the proper shape for a thin (uniform) arch. Truesdell notes that 'the problems of the catenary and the arch are reduced to one, but neither is solved'. It has been seen that LA HIRE in 1695 and GREGORY in 1697 both attacked the problem, from very different view points, of the stability of arches made from smooth (frictionless) voussoirs. Gregory was interested in the mathematical properties of the catenary; paradoxically, his mathematics can be criticized, while at the same time he seems to have had a deep grasp of the application of his work to the arch problem. By contrast La Hire, although he developed a powerful tool in the funicular polygon, made little headway in his 1695 *Traité de Mécanique*. The assumption of smooth voussoirs led him to an evident absurdity, and he could only conclude rather lamely that friction would confer stability on a structure which he had 'demonstrated' to be unstable.

In his *Mémoire* of 1712, however, La Hire made a considerable step forward. He observed that a weak arch would tend to crack at points about half-way between the springings and the crown. Although he did not describe these weak joints as hinges, he clearly saw that the line of thrust was constrained both in position and direction, being compelled to be tangential to the intrados at these points (cf. figs. 2.3 and 2.4 above). The hinge point being known, then the thrust in the arch can be calculated by statics, fig. 2.4, and the necessary width of the abutments to sustain that thrust can be determined.

La Hire did not discuss the question of whether the line of thrust could be contained within the masonry between the hinge points, but assumed that the crown section of the arch would be stable. In fact, for a circular arch (or indeed for any arch that is not very flat between the assumed hinge points), the thickness required to contain the thrust line is very small indeed. Thus if a circular arch of radius R and thickness t (fig. 6.5) carries its own weight only, and is assumed to form hinges at $\beta = 50°$, then the required ratio t/R is only about 4 per cent. Coulomb's critical comment (p. 40; p. 69) is relevant only if a very flat central section of the arch is supported by steep haunches.

PARENT, in the third volume of his 1713 *Essais*, devotes some attention to the problem of arches. The twelfth *Mémoire* in this volume (p. 152 ff.) discusses the arch of general shape and also specifically

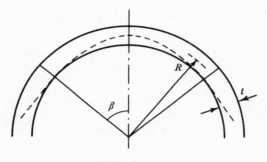

Fig. 6.5

the plate-bande, but the work cannot be said to add to that of La Hire. Similarly, GAUTIER's *Dissertation...* of 1717 makes no advance in the theory of arches, but has other matters of interest (in addition to the discussion of soil thrusts noted in chapter 5). He gives sets of tables for clear spans up to 120 ft, from which may be read the required thicknesses of the abutment piers and of the internal piers, as well as the dimensions of the voussoirs.

Further, Gautier reports what seem to be the first model tests on voussoir arches. Using nine wooden voussoirs he constructed half-arches between the ground and a vertical surface; these half arches were backed with further wooden blocks at the springing to prevent spread. The backing blocks were removed one by one until collapse of the arch occurred; the results of ten tests are given, each arch having its dimensions scaled from Gautier's tables. (Gautier also gives the results of some experiments concerned with the friction of ropes passing round fixed shafts. This again was one of the 'classic' problems of the eighteenth century, on which Coulomb himself wrote a *Mémoire*: 'Theorie des machines simples...', 1785. EULER had obtained in 1775 the result $T/T_0 = e^{\mu\theta}$.)

BÉLIDOR's contribution of 1729 in the *Science des Ingénieurs* (which was discussed above) is in some respects a step backwards from La Hire. La Hire left open the question of where the arch formed cracks, but Bélidor, in order to derive definite practical rules for design, assumes the section (for a semi-circular arch) to be at 45°. Further, Bélidor seems to have lost the idea of the hinge, since he places the thrust at this section not tangential to the intrados, but to the centre line of the arch.

Couplet

COUPLET's two *Mémoires,* of 1729 and 1730, are in fact a single paper divided into two parts:

Je divise ce Mémoire en deux parties. Dans la premiére, j'examine la forme & la poussée des Voûtes, & l'épaisseur de leurs pieds droits, sans y faire entrer l'engrénement des Voussoirs qui les empêchent de glisser les uns contre les autres.

Dans la seconde partie, je détermine les plus petites épaisseurs que l'on puisse donner aux Voûtes circulaires uniformes; j'en détermine aussi les poussées, en y faisant entrer l'engrénement & la liaison des Voussoirs qui les empêchent de glisser les uns contre les autres, ce qui n'a point encore été examiné par personne que je sçache.

Coulomb's *Remark I* (p. 38; p. 68) about the voussoirs slipping one on another is here anticipated by over 40 years, and Couplet seems satisfied that he is the first to introduce this idea. Indeed his first *Mémoire,* dealing with smooth voussoirs, contains no new concepts, but deals merely with different cases of loading. Couplet himself concludes the first *Mémoire.* by saying that the whole of the theory presented would be useless without an examination of the case when the voussoirs are prevented from sliding. The *Histoire* of the same volume (1729) is even more explicit in stating that any rules developed for arches must allow for the interaction of the voussoirs.

The 1730 *Histoire* frankly states that the frictionless hypothesis for the theory of arches is, in fact, wrong:

M. Couplet continuë la Théorie des Voûtes, qu'il n'avoit donnée en 1729, que dans l'hipothese purement géométrique & réellement fausse, que les Voussoirs fussent parfaitement polis. Ici il reprend la réalité, les Voussoirs s'engrénent par leurs surfaces les uns dans les autres, & il y faut même ajoûter ce qui n'est pas tout-à-fait réel, qu'ils s'engrénent de façon à ne pouvoir céder à aucune force, dont l'effet ne seroit que de faire glisser une surface sur une autre; car la Géométrie ne peut jamais s'allier à la Méchanique, qu'en y supposant quelque chose de plus absolu & de plus précis que le vrai.

Thus the material is assumed to have infinite frictional strength, and this assumption is known to be untrue. However, mathematics cannot come to the aid of engineering unless certain simplifying assumptions are made.

Couplet excuses his first, frictionless hypothesis by stating that all previous writers had made that assumption; secondly, he wishes to compare the values of thrusts resulting from the smooth hypothesis with those resulting from the frictional hypothesis. In his introduction to the 1730 *Mémoire*, Couplet is precise about his assumption that the voussoirs are sufficiently bound (*liés*) so that they cannot slide one on another. By this he does not mean that the arch is monolithic; on the contrary the voussoirs are merely locked together against sliding, and no resistance is offered to separation of the voussoirs during collapse

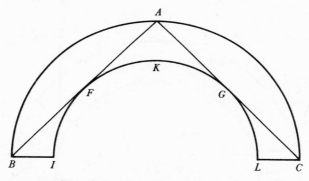

Fig. 6.6. After Couplet (1730).

of the arch. Thus Couplet made, explicitly and clearly, the three basic assumptions discussed earlier in this chapter: that stone has no tensile strength, that stone has infinite compressive strength, and that sliding failure cannot occur.

Couplet then states a *theorem*: if it is assumed that the voussoirs cannot slide one on another, the arch will not collapse if the chord of half the extrados does not cut the intrados, but lies within the thickness of the arch. Couplet's proof of this theorem contains precisely that aspect of duality of structural behaviour to which attention has been drawn. Whatever the load at the crown *A*, fig. 6.6, it can communicate directly with the abutment *B*, following the straight line *AFB* contained within the thickness of the arch. Further, says Couplet, for the arch to collapse the angle *BAC* must open, and this can only happen by a spread of the abutments, which is ruled out by hypothesis.

Couplet has in mind an arch of negligible self-weight subjected to a single point load at the crown *A*. Independently of the magnitude of the load a (straight) thrust line can be found which lies wholly within the arch. Further, there is no arrangement of hinges in the extrados and intrados which is at the same time compatible with a thrust line for the load and which gives rise to a mechanism of collapse. (The idea of a collapse mechanism is not discussed by Couplet at this point; as will be seen, however, he envisaged a mechanism as the failure mode.) Eccentric loading is not considered; as will be seen from fig. 6.7, a load acting away from the crown on the same arch will give rise to a mechanism (four-bar chain), since a complete thrust line cannot now be drawn to lie within the masonry.

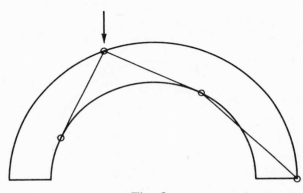

Fig. 6.7

The first problem stated by Couplet is to find the least thickness to be given to a uniform semi-circular arch, carrying its own weight only. The arch, says Couplet, will collapse by forming four pieces, attached to each other by hinges (*charniéres*), fig. 6.8. By considering the equilibrium of the arch in this state, a single equation can be found relating the thickness of the arch to its radius. Couplet obtains this equation (a cubic), and solves it numerically to obtain the required ratio of thickness to mean radius t/R as 10.1 per cent. There is only one trivial error in this analysis; Couplet does not determine the proper position of the hinges *T* and *K* at the intrados, but assumes, with Bélidor, that they form at 45°. Thus in considering the equilibrium of the piece *AK* of the arch, the thrust at *A* must be horizontal and may be represented by *AG*, while the weight acts through *H* and is represented by *GH*; the thrust at *K* must therefore act in the line *KG*, which is not tangential to the intrados at *K*.

172

Couplet misses this point, but his analysis is otherwise completely valid.

As Coulomb noted (p. 39; p. 69), the calculations are insensitive to the precise position of the hinge at the intrados. Figure 6.9(a) shows (approximately to scale) the thinnest semi-circular arch which will just contain a thrust line due to its own weight. The angular location β of the hinges at the intrados is given by the solution of the equation

$$\beta \cot \beta \left[\frac{2\beta \cos \beta + \sin \beta \cos^2 \beta + \sin \beta}{2\beta \cos \beta + \sin \beta \cos^2 \beta - \sin \beta \cos \beta} \right] = \frac{\pi}{2}. \qquad (6.1)$$

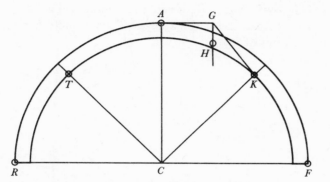

Fig. 6.8. After Couplet (1730).

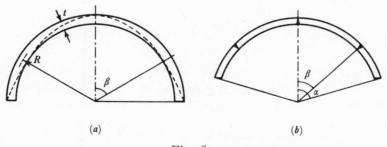

(a)   (b)

Fig. 6.9

Having found this value of β, the required thickness ratio t/R is then calculable from

$$\frac{t}{R} = 2 \frac{(\beta - \sin \beta)(1 - \cos \beta)}{\beta(1 + \cos \beta)} \qquad (6.2)$$

Equation (6.1) gives $\beta = 58° \; 49'$, and, from equation (6.2), $t/R = 0.106$ (rather than the 0.101 found by Couplet).

For the incomplete circular arch of angle of embrace 2α, fig. 6.9(b), equation (6.2) is unchanged, but the value of β is determined from a new equation (6.1), in which the right-hand side, $\frac{1}{2}\pi$, is replaced by $\alpha\cot\frac{1}{2}\alpha$. Couplet calculates in his second problem the required ratio t/R for an arch embracing 120° rather than 180°. He assumes the same collapse mechanism, with the arch breaking into four equal pieces, and finds t/R to be 1·95 per cent. The correct value is 2·26 per cent, with the angular position of the hinges at the intrados about 41° rather than the assumed 30°. (For α small, the solution of the revised equation (6.1) is approximately $\beta = \alpha/\sqrt{2}$, and in fact the ratio β/α lies between 0·71 and 0·65 over the whole range $0 < \alpha < \frac{1}{2}\pi$. Figure 6.10 shows the least thicknesses of incomplete circular arches for various values of α.)

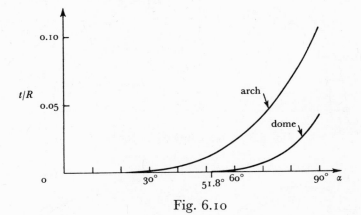

Fig. 6.10

Couplet's third and fourth problems are concerned with the value of the arch thrust and with the consequent design of the supporting piers; the calculations are straightforward, and will not be discussed here.

This contribution of Couplet is remarkable. He obtained an essentially complete and correct solution to the problem of arch design, he had clear notions of lines of thrust and of mechanisms of collapse caused by the formation of hinges, and he stated explicitly the simplifying assumptions necessary for the analysis. Coulomb's concluding remarks (pp. 39–40; p. 69) refer to La Hire and to Bélidor only, and he notes that these authors divide the arch, for purposes of analysis, into three pieces. He states that a division into four pieces would often give a more severe case. This division into four pieces was precisely one of Couplet's contributions; what was required was experimental confirmation, and this was provided two years after Couplet's second *Mémoire*, by DANYZY in 1732.

Eighteenth-century work (2)

FRÉZIER, in the twelfth chapter of the third volume of his treatise on stereotomy, published in 1737, gives appendices on the determination of arch thrusts, and hence of the dimensions of the piers. He reviews the work of La Hire and of Couplet, noting that this is theoretical, and he then comments on the experimental results of DANYZY. These results were presented to the *Société Royale* of Montpellier on 27 February 1732, but the results remained unpublished (except by Frézier) until 1778. Danyzy tested arches made of small plaster voussoirs, and Frézier's figs. 235–40 (fig. 6.11 here) record some of the experimental results. (Figure 241 is redrawn from Couplet, cf. fig. 6.6 here; the other figures reproduce almost exactly those given by Danyzy in 1778, but are rather better drawn.) All the arches shown are on the point of collapse, the piers having their minimum dimensions. Figure 235, for example, records a collapse mode exactly predicted by Couplet (apart from the double hinge at the keystone). The tangency of the thrust line to the intrados is clearly 'visible' in, for example, fig. 238. Figure 240 is of interest; the plate-bande can collapse only if the abutments spread.

Thus by about 1740 the mechanics of the arch was well understood, and the whole theory could be applied to the analysis and design of masonry. Apart from one such application, however, by POLENI (which is discussed below), it would seem that the work was gradually forgotten from about 1750 onwards. Coulomb did not refer to Couplet, and in France the theory was once again rediscovered by LAMÉ and CLAPEYRON in 1823, as will be seen; there followed further periods of forgetting and rediscovery.

In Italy, however, the *tre mattematici*, LE SEUR, JACQUIER, and BOSCOVICH, two Minims and one Jesuit, were commissioned to report on the state of the dome of St Peter's; this they did in 1743. Cracking had caused alarm, and the investigators concluded that further ties were needed at the base of the dome to contain the thrust. Their calculations included the postulation of a collapse mechanism involving hinges, and an assessment was made of the value of the tie force by using the equation of virtual work. The report met with severe criticism, mainly because it was felt that mathematics was not the proper tool to use in the analysis. In 1743 POLENI was also

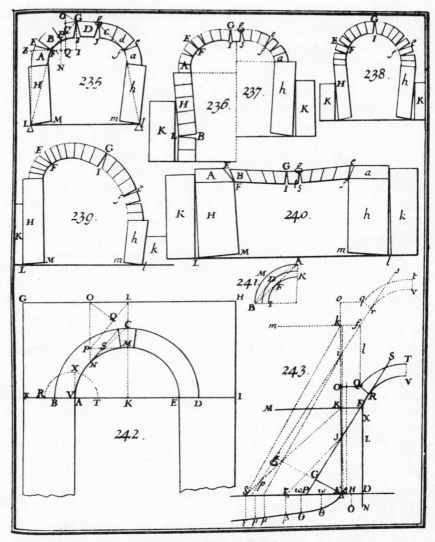

Fig. 6.11. From Frézier.

commissioned to report on the dome, and he published his *Memorie istoriche* in 1748.

Poleni gives a comprehensive review of the existing state of knowledge of masonry construction, including the French work (La Hire, Parent, Couplet etc.). He also quotes Gregory, and an interesting development of Gregory's work by STIRLING. (Stirling's book of

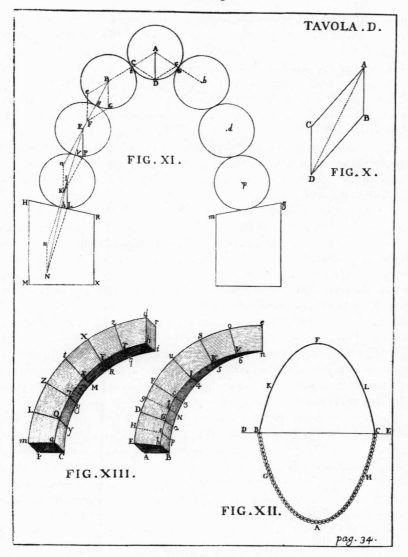

Fig. 6.12. From Poleni.

1717 has, as a sort of *additamentum*, a brief discussion of the inverted catenary formed by the balancing of smooth spheres. Poleni has an illustration, fig. 6.12, closely based upon that of Stirling.) Thus Poleni had on the one hand the idea of the shape of a frictionless arch, and on the other he knew of the work of Couplet on real arches, in which failure occurs by the formation of hinges. Echoing Gregory, he then states that in order for the arch to be stable, all that is necessary is that the line of thrust should everywhere lie within the masonry: 'che dentro alla solidità della volta la nostra catenaria tutta intiera sia situata'; and again, stating the principle the other way round: 'E per dir brieve, in questo esame fatto colla Catenaria, il punto principale consistera nel vedere, se veramente alcuna parte della Catenaria cadesse fuori de' contorni della Volta.'

All the work up to that of Poleni had been done on two-dimensional problems; while papers were written nominally on *voûtes* or *volte*, the mathematics dealt with the problem of the arch (as defined by Coulomb, pp. 28ff; p. 61). Poleni applied the theory to the analysis of the three-dimensional dome. The cracks at St Peter's had already divided the dome into portions approximating half spherical lunes (orange slices); for the purpose of his analysis, Poleni sliced the dome into 50 such lunes, of which one is shown in the right-hand drawing of his fig. XIII, fig. 6.12. He stated that if each lune would stand, then so also would the complete dome. The thrust line was determined experimentally by loading a flexible string with unequal weights, each weight being proportional to that of a segment of the lune, and due allowance being made for the weight of the lantern. Poleni's experimental result is shown in fig. 6.13, and the thrust line found in this way does in fact lie within the thickness of the dome of St Peter's; the figure also shows that a uniformly loaded string would produce an equivalent thrust line passing outside the masonry.

Poleni concluded that the observed cracking was not critical, but he agreed with the *tre mattematici* that further ties should be provided. It will be seen from fig. 6.13 that the dome is solid only for about 20° from the springing; thereafter the dome splits into two shells between which one may climb to the lantern. In reaching his conclusion, Poleni was not disturbed by the fact that his experimentally-derived thrust line fell partly in the void between the shells; this is but one instance of his deep understanding of the problem. Indeed, Poleni's attack is one that would be almost exactly reproduced by a modern analyst using the safe theorem of plasticity. Admissible thrust lines for the quasi two-dimensional lunes are clearly admissible for the original structure; if the sliced dome will stand, then so *a fortiori* will the complete dome.

As a matter of interest, the orange-slice technique may be used to obtain minimum thicknesses for spherical domes (Heyman 1967) exactly as for the two-dimensional arch. Figure 6.14 (approximately

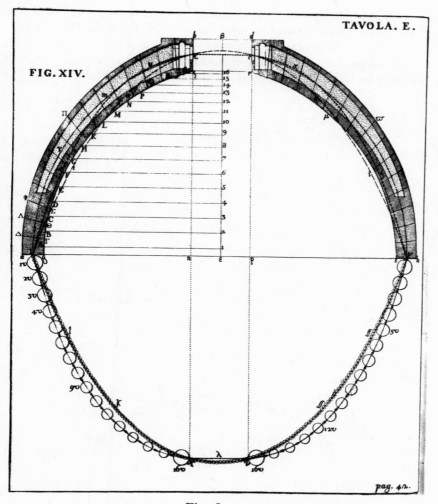

Fig. 6.13

to scale) shows the limiting position of the thrust line for a very thin
lune; the thrust line touches the extrados of the lune at *P* and the
intrados at *Q*. The corresponding collapse mechanism for the complete dome is shown in fig. 6.15, in which a central portion near the
crown does not deform, but merely descends vertically. (It is this
which explains the great flexibility in the design of practical domes;
the crown can be completely omitted, i.e. the dome can have an 'eye',
or, alternatively, a heavy lantern may be loaded on to the crown.)

As the dome collapses, adjacent lunes will separate between P and the base hinge; such separation can occur quite freely in the 'sliced' structure, and also actually if the masonry is assumed to be incapable of taking tension. (It was the thrust H in fig. 6.14 for which Poleni and the *tre mattematici* required extra ties to be provided.)

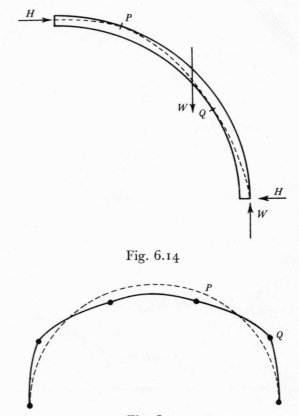

Fig. 6.14

Fig. 6.15

The minimum thickness of the dome in fig. 6.14 is 4.20 per cent of the radius. The analysis may be repeated for incomplete domes of angle of embrace 2α (as for the arch of fig. 6.9(*b*)), and the results are shown in fig. 6.10. As α is reduced, so is the required thickness, until at $\alpha = 51.8°$ (actually $\cos\alpha = \frac{1}{2}(\sqrt{5} - 1)$) the hinge points ($P$ and Q in fig. 6.14) coincide, and the thickness falls to zero. The significance of this result is of course that the stress resultants for a uniform spherical shell give purely compressive stresses providing α is less than $51.8°$; thus, in theory, a spherical compressive thrust *surface* may be found to coincide with the shell over this portion of the dome.

Coulomb and Bossut

It is not absolutely safe to say that Coulomb had not heard of Poleni's work, but he betrays no knowledge of it in the *Essai*; certainly his training seems to have been based on Bélidor, who mentions La Hire but stops short of Couplet, let alone Poleni. Coulomb starts in the traditional way by considering (pp. 29–34; pp. 61–6) the behaviour of frictionless arches, but he himself remarks, both in the text and in the introduction, that the theory is only of limited use. He then proposes the sliding mode of failure (p. 35; p. 66), allowing for friction and cohesion, and finally, by inhibiting the sliding mode, he arrives at the notion of collapse by the formation of hinges (p. 38; p. 68). In this presentation he is to some extent covering the ground already trodden by Couplet, but he obtains his results more rigorously. Coulomb is very precise in the formulation of his assumptions and of the conditions which any solution must satisfy. These formulations enabled him to discuss the problem of bending of a beam with economy and clarity; similarly his solution for the arch does not contain Couplet's (trivial) error, since he allows the hinge position at the intrados to be determined by a variational process, rather than assuming it to occur at a fixed point ($45°$ for a semi-circular arch).

In this work, Coulomb's Preliminary Proposition I (p. 4; p. 43) is prominent, and he refers continually to his first and second conditions of equilibrium. If the voussoirs cannot slide on each other, then the first condition can be ignored (p. 38; p. 68), and Coulomb deduces the single condition necessary to confirm the stability of an arch, namely that the line of thrust should lie everywhere within the masonry. This is the conclusion of Coulomb's work on arches; it is a powerful conclusion, and Coulomb backs it by mathematical argument, but he has in fact made no real advance on the work of Poleni a quarter of a century earlier.

A *Mémoire* on arches by Coulomb's teacher, the Abbé BOSSUT, appeared in the proceedings of the *Académie* for 1774, a year later than the nominal date of the *Essai* (which was read in 1773, but not published until 1776). A footnote to Bossut's *Mémoire* states that it was read in 1770 (12 July) and that the author's intention had been to subjoin some results on the strength of stone; however, he had had no chance to make the necessary tests, and now wished to publish the theory. Although appearing in the 1774 volume, the final manuscript was received by the *Académie* on 5 September 1777, and published in 1778. Bossut gives a brief historical introduction, mentioning among others Bernoulli, La Hire and Couplet; he adds that all other works that he knows reduce essentially to those mentioned. To this last statement is added '*En* 1770'.

Bossut's *Mémoire* is, in fact, a 'pre-Coulomb' paper, written before 1773 but published when the author knew of Coulomb's findings. (Bossut officially examined the *Essai*, and reported to the *Académie* on 30 April 1774.) The first section deals mainly with frictionless voussoirs; Coulomb covers very much the same ground in his pp. 29–30; pp. 61–2. Bossut then has a brief analysis of La Hire's hypothesis that the arch breaks into three pieces, but he adds very little; the rupture joints at the intrados are left undefined.

The second portion of Bossut's *Mémoire*, of only 11 pages, is of much greater interest, since the equilibrium of the dome of revolution is considered. Bossut mentions only one previous paper on this subject, which he says has nothing in common with his own work; he does not know of Poleni's study. However, he slices the dome into lunes, exactly as Poleni had done, but instead of considering lines of thrust, Bossut applies La Hire's theory, treating each segment of the dome exactly as an arch. The work is thus open to Coulomb's criticism (p. 40; p. 69), particularly since Bossut allows for the weight of a heavy lantern, and does not demonstrate that a proper line thrust can be found. The work is nevertheless intuitively correct, and Bossut applies it to show that Soufflot's dome to the Church of St Geneviève in Paris was indeed stable. He assumes a certain location for the rupture points in this analysis, finally fixing them by noting that a small change in position, either up or down, has no effect on the final result. This is, of course, Coulomb's method of Remark II (pp. 38ff; p. 69), but there is no indication of whether Bossut had got this variational principle from Coulomb, or *vice versa*, or whether both had arrived at it independently.

Bossut read a further short paper on domes on 2 September 1778, and this is published in the *Histoire de l'Académie* for 1776, printed in 1779. These 10 pages amplify the previous *Mémoire*, and state clearly that the sliced segment may be treated exactly as the parallel arch for the purpose of the stability analysis. Bossut still uses the approach of La Hire; that is, he obtains a solution which is derived neither from a proper mechanism of collapse nor from a proper (safe) thrust line, but which is, nevertheless, 'reasonable'.

There seems to be little further theoretical work in the eighteenth century, either on domes or on arches, but it may be noted that great advances were being made in the practical art of bridge building at the time that Coulomb and Bossut were writing. PERRONET had flattened the arch and reduced the dimensions of internal piers; his Pont de Neuilly was built in 1768–74. Perronet's influence was very great, both as a designer of major works, and as the first director of the newly-formed *École des Ponts et Chaussées*, which will be mentioned again in chapter 7; he believed particularly in the use of applied research as an aid to the solution of engineering problems.

The nineteenth century and later

BOISTARD made a very careful series of tests on model voussoir arches in 1800; his paper is contained in a volume collected by LESAGE in 1810 *à l'usage de MM. les ingénieurs (des Ponts et Chaussées)*. Boistard is not very concerned with the *theory* of arches, and gives only brief calculations at the end of the paper. In the introduction he mentions only Prony and Couplet; he states that Couplet's hypothesis of the mode of collapse is essentially correct, but that the rupture points should not be fixed midway round the intrados, where Couplet has placed them. The model tests had a threefold object; first, Boistard wished to obtain the precise modes of collapse under various loading conditions, including superimposed loading; second, he wished to determine minimum abutment requirements under these conditions, that is, to investigate the by now standard problem of vault thrust; and third, he wished to obtain some idea of forces on the centering during construction of an arch.

This last problem, of obvious practical importance, does not seem to have been discussed before. In Boistard's first test, for example, the semi-circular arch of 48 voussoirs was left for several days on the centering before the keystone was inserted. He observed that, under the conditions, the arch was not in contact with the centering from the springings up to the eleventh voussoir on each side. This ability of the springing voussoirs to carry themselves was, of course, exploited by the Romans; see, for example, FITCHEN's account of the probable constructional technique used for the Pont du Gard. Boistard, however, merely records his experimental findings, and neither calculates nor speculates on these.

Boistard's paper consists of the results of 22 tests on arches of various shapes. He developed an elegant and economical way of making voussoirs, each of 4 in thickness, from two polished bricks; in all cases, the voussoirs were assembled to form an arch of 8 ft clear span. Thus in the first test on a semi-circular arch the ratio t/R was $4/50 = 8$ per cent; Couplet had predicted that the minimum ratio for stability was 10.1 per cent (actually 10.6 per cent), and Boistard found indeed that the arch collapsed when an attempt was made to decenter it. The collapse mechanism is recorded.

There are 4 tests on semi-circular arches, 11 on oval arches (*voûtes surbaissées*), and 5 on small-rise circular arches of height to span ratio $\frac{1}{4}$ and $\frac{1}{8}$. Test 21 had a height-to-span ratio of about $\frac{1}{16}$, and was a model (to a scale of $4/25$) of the very shallow Pont de Nemours. The final test was on a plate-bande of the same span of 8 ft, which failed by overturning of the abutments (cf. Danyzy's fig. 240; fig. 6.11). Boistard's paper concludes with a brief theoretical discussion (some of it wrong), with some test results on mortar (the experiments had actually been made on dry voussoir arches), and with some notes on the problems of decentering.

Lesage's collection of papers on bridges was drawn from the *Bibliothèque Impériale des Ponts et Chaussées*. GAUTHEY, who died in 1807 and who had been Perronet's pupil, was *Inspecteur-Général des Ponts et Chaussées*; his three-volume *Traité de la construction des ponts*, edited by Navier (his nephew), started to appear in 1809. This treatise assembles and digests all the theoretical and experimental work known to the *Ponts et Chaussées* by the beginning of the nineteenth century; the treatise is at once a history of bridge-building, a survey of existing bridges, an architectural hand-book and, above all, a manual on the design and construction of masonry arches, together with their specification and costing. Further research was certainly needed, but the art had reached the stage where the production of such a text-book and code of practice was possible. By contrast, a work such as Rondelet's *l'Art de Bâtir*, whose fifth edition was produced in 1812, must have begun to look very old-fashioned.

In Gauthey's treatise the research (or scientific) elements of the behaviour of masonry tend to be hidden in the practical rules; the treatise itself would not be likely to give rise to further research. Thus further work tended to go back to the findings of La Hire or Couplet, and to be published for the 'scientific' rather than the 'engineering' world. Perhaps the supreme example of this is the long (over 300 pages) *Mémoire* of YVON VILLARCEAU, presented to the *Académie* in 1845 and published in 1854. Yvon Villarceau knew that a given arch was essentially statically indeterminate, and that there therefore existed an infinite number of possible equilibrium states; he developed a 'safe' design method by requiring the centre line of the arch to coincide with one of the possible thrust lines for the given loading. This inverse design method requires the numerical solution of the equations, and the results are presented in the form of tables which can be used immediately in standard calculations by the bridge designer. As a design method there is really nothing to be added; Yvon Villarceau's work could be used today with confidence and economy.

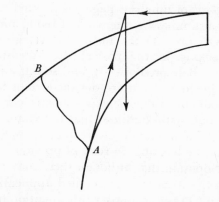

Fig. 6.16. After Lamé and Clapeyron.

The process of rediscovery, however, continued. LAMÉ and CLAPEYRON, in one of their first assignments on their secondment to Russia, had to go back to first principles to assess the stability of the dome of the cathedral of St Isaac in St Petersburg. Without mentioning either Couplet or Boistard, they state that the only tenable theory of failure of arches is that obtained by the study of a hinged mechanism. Similarly, without mentioning Coulomb, they state his variational principle for finding the fracture joint at the intrados. Using these ideas, they show that the hinge position is given by the condition that the tangent to the intrados at that point, the line of action of the weight of the half arch lying above that point, and the horizontal through the extrados at the crown of the arch, should all be concurrent; the triangulation of forces will be apparent from fig. 6.16. This theorem gives a quick graphical method for estimating the value of the arch thrust, although some assumption

13-2

has to be made as to the line of fracture *AB* at the critical section (if this is not defined by joints between voussoirs).

Lamé and Clapeyron then apply their theory to the cathedral of St Isaac. They observe that the dome is supported on a cylindrical wall, pierced by 12 arcaded portals, so that there are 12 weak meridional planes, and they propose to establish the stability of the whole dome by establishing the stability of a demi-lune of 30°. Thus they recreate Poleni's slicing technique, and treat the lune as an arch.

The *Mémoire* was published in 1823 in the *Annales de Mines*, but was reported on in the same year to the *Académie* by Dupin and Prony; their report, of nineteen pages, is printed with the *Mémoire*. In the main it is a summary of the work of Lamé and Clapeyron, but the *rapporteurs* also provide the historical background which is so conspicuously lacking in the paper; they mention the work of La Hire, Couplet, Bélidor, Boistard, and so on, and conclude 'que ces deux jeunes ingénieurs ont été devancés dans la découverte des bases fondamentales de la théorie exposée dans leur mémoire'. However, they point out, with kindness, that the young men had hardly finished their studies when they were posted to Russia, a country where they would be less able to read of previous work. Dupin and Prony are of the opinion that although there may be nothing fundamentally new in the *Mémoire*, the detailed application of the theory is of great merit. Prony cannot resist mentioning that his own *Mécanique philosophique* of 1800 gave the equilibrium equation of an arch in exactly Lamé and Clapeyron's form.

In due course NAVIER published his own *Leçons* at the *École des Ponts et Chaussées*, of which the third edition was annotated at such length by Saint-Venant. In the second edition about fifty pages are devoted to arch theory (not reproduced in the third edition). This section is the scientific counterpart of Gauthey's treatise; it has the same awareness of the historical development of the subject (including references to Poleni and the *tre mattematici*), but Navier, unlike Gauthey, is concerned with detailed exposition of the theory. The presentation in fact follows very closely that of Coulomb, but Navier is explicit about conditions for stability. Coulomb had stated that if an arch was stable, then the thrust must lie between certain calculable limits. Navier states the converse of this, which is the equivalent of the safe theorem; if the thrust lay between the limits, then this was a proof of the stability of the arch.

As examples of the intuitive appreciation of the safe theorem, taken almost at random from the mid-nineteenth century literature, the following quotations come from a standard text-book in use at the Royal Military Academy in England. The author, J. F. HEATHER, who was for a long time a master at the Academy, deals with arches in 10 pages (out of a total of some 360 on mechanics). Heather gives diagrams of collapse mechanisms and of corresponding thrust lines, but quotes no sources. Having defined the parts of an arch, he states:

The condition for the stability of an arch is that it...contain within its thickness one line of pressure at least, under the most disadvantageous circumstances in which it can be placed.

If the surfaces of the voussoirs were perfectly smooth, the voussoirs would slide past one another, unless the line of pressure were everywhere perpendicular to the faces of the joints, but in the forms of arches usually adopted, the direction of pressure at any joint cannot, while the above condition is satisfied, make with a perpendicular to the face of a joint an angle greater than the limiting angle of friction, and consequently no sliding could take place even if the arch were uncemented.

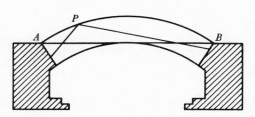

Fig. 6.17. After Heather.

Later, Heather has a *proposition* which is reminiscent of Couplet's theorem: 'PROP. If the tangent to the highest point of the intrados when produced to the abutments lies wholly within the thickness of an arch, the arch will bear any weight which does not crush the material of which it is composed.' (Figure 6.17.)

It would seem that Heather is expounding well-known facts, from the way in which he writes, but the ideas cannot have been current for very long. John Weale had published in 1843 two volumes on bridges, of which the first, under the editorship of JAMES HANN, contains translations from Gauthey, and also a long paper by Moseley on the theory of the arch. MOSELEY's own *Engineering and Architecture* was first published in 1843 also; he states in the preface that, when he worked out his theory of arches, that of Coulomb was not known to him. Moseley himself owed a great deal to his reading of French authors, to whom he makes acknowledgement, and in particular to the work of Navier and of PONCELET, whose *Mécanique Industrielle* was published in 1839. Although his arch theory was developed independently of that of Coulomb, the conclusions are the same, and Moseley's work, although influential at the time, appears now to be one more instance of the constant rediscovery of the principles of the action of masonry.

The masonry arch was, of course, already obsolescent by the mid-nineteenth century (SÉJOURNÉ published in 6 volumes in 1913 a definitive catalogue of large-span masonry bridges throughout the world, complete with drawings and calculations and details of design and construction). Rennie's new London Bridge, now taken down, was completed in 1831, and Thomas Harrison's bridge at Chester, at 200 ft the largest masonry span in England, was built a year later.

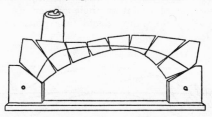

Fig. 6.18. From Ewing.

These were among the last of the masonry bridges. Iron Bridge at Coalbrookdale had been built in 1779; Telford had already projected a cast-iron span of 600 ft for the new London Bridge, and the future lay with Stephenson, Brunel, and Benjamin Baker.

FLEEMING JENKIN's long article on bridges for the ninth edition (1876) of the *Encyclopaedia Britannica* devotes most of the space to wrought iron and steel, and very little to masonry. In his discussion of voussoir arches, however, Fleeming Jenkin is well aware of the safe theorem, and he made an ingenious model (fig. 6.18) to give an 'Experimental Demonstration that the Equilibrium of a series of Voussoirs is stable if any Equilibrated Polygon can be drawn fulfilling the conditions stated above' (that the thrust line should lie within the arch). The voussoirs in the model had curved faces so that they rocked on each other; the line of thrust corresponding to their own weights and to any external loads could then be 'seen' by observing the points of contact of the voussoirs. (EWING reproduced Fleeming Jenkin's sketches in his *Strength of Materials* of 1899. It is in this book that the first modern rediscovery of the full plastic moment is made; Ewing derives the elastic and plastic moduli for a rectangular beam, $\frac{1}{6}bd^2$ and $\frac{1}{4}bd^2$, and he discusses the case of partial plasticity.)

There was a renewed interest in voussoir arches just before the Second World War, due to the work of PIPPARD *et al.* Pippard made careful tests of model arches with steel voussoirs, and demonstrated that the slightest imperfection of fit (e.g. at the abutments) converted an apparently redundant into a statically determinate structure. Although his experiments were different, Boistard had covered much the same ground in 1800; in the intervening century, however, the whole idea of statical indeterminacy had been developed. Pippard interpreted his results by reference to principles of minimum elastic energy; it was not yet time for the further rediscovery of the limit principles of plasticity.

Scholium

In the whole of this work, from La Hire onwards, there is almost no reference to the strength of the material used in arch construction. There is an occasional warning that the line of thrust must not approach too close to the surface of the masonry, or the edges of the voussoirs might crush; or again, Perronet proposes that internal piers, carrying vertical load only, may be reduced in size until the working limit of the material is reached (whereas external abutment piers must be proportioned against overturning). In general, however, full use was made of the fact that stress levels in this type of construction are extraordinarily low. Thus the work of the seventeenth and eighteenth centuries was directed towards finding the *shape* of an arch in order that it should be stable, and, in this, there is a direct continuation of the traditions of the Middle Ages. The 'secrets' of the mediaeval masonic lodges turn out to be numerical, and relate the proportions that one part of a structure should bear to any other part (see e.g. FRANKL). Such rules are essentially correct if material properties play no part in the stability of a structure. The square–cube law does not apply; if a model of a voussoir arch stands, then so also will the full size structure, and it is possible to define 'geometrical' factors of safety rather than safety factors on stress (Heyman 1968, 1971).

It was this mediaeval tradition that Galileo warned Sagredo against in his *Due nuove scienze* (see p. 90 above); in a sense, the history of the theory of arches is a history of the attempts of engineers to escape from Galilean concepts of strength of materials. Not until the span of an arch is a clear 1000 ft (see BAXTER, GEE and JAMES on Gladesville Bridge) does any question of stress start to arise.

7

Some Historical Notes

Coulomb died on 23 August 1806; the 'Éloge historique de M. Coulomb' was published by the *Institut* that same year, and read publicly by its author DELAMBRE, on 5 January 1807. Another eulogy, by BIOT, was included in the *Biographie universelle* of 1813, and later reprinted. All subsequent published notes on the life of Coulomb can be traced back to one or other of these two authors (e.g. HAMILTON 1936–7, HOLLISTER 1936, TIMOSHENKO 1953), while POTIER's preface to the collected edition of 1884 of Coulomb's *Mémoires* (which does not, however, contain the *Essai*) is based directly on the official *Éloge*. An exception may perhaps be found in Thomas YOUNG's article on Coulomb for the 1824 Supplement to the *Encyclopaedia Britannica* (reprinted in his collected works). Young gives, for example, a very good summary of the *Essai*, which lasted through two editions and fifty years, and was only suppressed in the 9th edition of 1876. Here Coulomb is allotted only half a column rather than 6 double pages; from the 11th edition of 1910 onwards all mention of the *Essai* has disappeared.

There seems to have been no direct archival work on Coulomb until that of GILLMOR, whose biography was published in 1971, and which gives for the first time a connected account of Coulomb's development as a military engineer and as a scientist. Further, Gillmor clears up several questions of fact, about which other biographers had been mistaken. As a trivial example, it appears that the Coulomb family was not noble, and there is no justification for using the style de Coulomb to imply nobility. Similarly, although Coulomb left Paris after the closing of the *Académie* in 1793, and retired to his property of 50 acres at Blois, there is no evidence that he was pursued by the revolutionary law prohibiting nobles from remaining in Paris.

Again, Hollister states that Coulomb's '... family had for generations been charged with the care of the fountains of France'. This statement appears to be based on a misreading of Delambre, who certainly mentions that, at the time of the Revolution, Coulomb resigned from *l'intendance générale des fontaines de France* (in 1792);

190

Delambre then adds that this post had been hereditary in a family which had become extinguished. However, the family was not Coulomb's, but, as Gillmor found, that of the comtes de Villepreux; the title *Surintendant des Eaux* had been bestowed by Louis XIII in 1623, and the last *comte* died in 1783.

Coulomb was appointed *Intendant des Eaux et Fontaines du Roi* in 1784 (with hereditary rights). The post sounds faintly ludicrous, but was in fact far from being a sinecure; the *intendant* was responsible for the supply of water to all the royal buildings in Paris, and to Versailles and Fontainebleau, as well as for public supplies tapped from the King's aqueducts, and there was some cooperation between the royal *intendant* and the administrators of the public supply to Paris. In fact, Coulomb had for some years before his appointment made a study of the problems of the pumping of water, and it was these studies that led to his nomination.

Delambre's official *Éloge* is silent on these matters, and is, indeed, reticent about much that is of present interest. Coulomb is remembered as a great scientist, whose fame rests on his discoveries in electricity and magnetism, on his invention of the torsion balance, and the consequent verification of the inverse square law for static electrical charges. He left to his sons (in the words of the *Éloge*) no other inheritance than a respected name, the example of his virtues, and the memory of the brilliant services he had rendered to science. His work on friction is mentioned only briefly. The *Essai* is neither named nor dated, but Delambre notes that, after his return from Martinique, Coulomb read a *Mémoire* to the *Académie* which earned him the title of *correspondant*.

Technical education in France

Coulomb was, in fact, an engineer by training, and had graduated from the engineering school at Mézières. The *Corps du Génie* had been organized in 1676, but without any school or other precise way of selecting its officers. A royal examiner was appointed in 1702 to make this selection, and in 1748 the academician Camus (Charles-Étienne-Louis, 1699–1768) succeeded to the post. The *École du Corps Royale du Génie* was established at the same time at Mézières, and opened its doors on 16 June 1749. The new school rapidly settled down into a routine which lasted until a reorganization in 1777. In 1750 Camus held the first graduating examinations, and 8 engineers were promoted from the school. By 1751 no pupil was admitted to the school unless he had been examined by Camus, and, in order to present themselves for this entrance examination, candidates had first to submit a drawing and undergo a *viva voce* in mathematics.

Courses at Mézières lasted two years, and included instruction in stereotomy, the making of maps, design of fortifications, 'et toutes les pratiques générales dont un ingénieur a besoin'. Camus insisted that all engineers should speak the same mathematical language, and constructed unified courses in arithmetic, geometry, mechanics, and hydraulics. He himself wrote texts for the first three subjects; those of Varignon and Mariotte were used for hydraulics until 1769, at which time Bossut supplied the standard text. Camus also made it quite explicit that bourgeois youths were to be admitted to the school as well as nobles.

Thus the *École du Génie* was unusual in having certain entrance requirements beyond the ability to read and write (and even these requirements were often waived at other schools). In 1756 entry to Mézières was made even more difficult; candidates had to have first attended the newly organized school at La Fère. (An artillery school had been in existence at La Fère since 1720, and Bélidor started teaching there at that date.)

The *École du Génie* became, in fact, a graduate school, and this was a pattern to be repeated in other branches of the profession. Indeed, after the establishment of the *École Polytechnique* in 1794, engineering students had a common basic training in Paris, and then went on to the specialist schools, such as the *École d'Artillerie* at Metz (after 1802 the *École du Génie* and the *École d'Artillerie* were combined at Metz), the *École des Mines*, the *École des Ponts et Chaussées*, and so on, all these higher schools being *Écoles d'Application*. As ARTZ remarks, 'by the early nineteenth century France was the only country in the world where engineering was clearly and definitely established as a learned profession'.

The *École Polytechnique* gave a three-year basic course, and had teachers such as Lagrange, Laplace, Prony, and Monge, men of the highest ability who were freed from routine work by a staff of assistants. The eminence of such men is in curious contrast to the appointment forty years earlier (in 1752) of the then unknown young man, the Abbé Bossut, to the professorship of mathematics at Mézières. This proved to be a key nomination by Camus, although it was not to the liking of the school's *commandant*, Chastillon, *ingénieur en chef* at Mézières. Indeed, five years later Chastillon made some criticisms of the school, and of Bossut in particular, as being too theoretical; these criticisms fit exactly into the pattern of a practical man's distrust of the use of applied mathematics. His view of Bossut was of a man who had no feeling for engineering, but who wished to make his name in pure mathematics, and, as such, was not fitted for his part at the school; moreover, Bossut was lazy.

It is probably true that Bossut did wish to make his career as a mathematician, and to advance himself up the well-defined ladder of the *Académie*. This was of course no unworthy ambition (it was a career to be followed subsequently by Coulomb), and it certainly did not necessarily unfit Bossut as a teacher in an engineering school; Coulomb's admiration of Bossut is strong evidence on the other side. In fact Bossut remained as professor at Mézières for 16 years until 1768, in which year he was both elected to membership of the *Académie* and succeeded Camus as *examinateur* at the school.

Bossut had been elected a *correspondant* of the *Académie* in 1752, at the age of 22, at the same time as his appointment at Mézières; he advanced to *adjoint* in 1768, to *associé* in 1770, and finally to *pensionnaire* in 1779, all in the class of *géométrie*. In the reorganization of 1785 he was renamed *pensionnaire de la classe de mécanique*, and, after the Revolution, he was *Membre de la première classe de l'Institut de France, section de mathématiques* (9 December 1795). As *premier pensionnaire* in the class of *mécanique* he received, in 1785, a *pension* of 3000 *livres* annually; the stipends of the second and third *pensionnaires* (Rochon and La Place) were 1800 and 1200 *livres* respectively.

(MAINDRON gives the following membership of the *Académie* after the reorganization of 1785:

12 *honoraires*
24 *pensionnaires*
24 *associés ordinaires*
 1 *associé géographe*
12 *associés libres*
 8 *associés étrangers*
 1 *secrétaire perpétuel*
 1 *trésorier perpétuel*,

compared with the original membership of :

10 *honoraires*
20 *pensionnaires*
20 *associés*
20 *élèves*.

In addition, *associés* and *pensionnaires* could have *correspondants*; Bossut was originally the *correspondant* of d'Alembert.)

Coulomb's early life

Coulomb was born on 14 June 1736, at Angoulême, but the family had lived for several generations at Montpellier. His father moved to Paris when Coulomb was a boy, and it is likely that, between the ages of 10 and 15, he received some mathematical instruction at the *Collège Mazarin*; he also attended some of Le Monnier's lectures on mathematics at the *Collège Royal de France,* and he decided that he wished to become a mathematician. However, his father returned to Montpellier, and Coulomb was forced to join him there for financial reasons.

Gillmor could find nothing of Coulomb's further schooling, if any, in Montpellier. Since 1706 there had been a *Société Royale des Sciences de Montpellier*, and, although he was at first too young to join, he certainly knew the members of the society. In 1757 he was admitted as *membre adjoint*, and in the same year he read a mathematical paper to a meeting held in the private house of Danyzy. This was the same Danyzy who, in 1732, had read (to the same society) the paper on arch thrusts noted in chapter 6. It was seen that the theory and the diagrams (fig. 6.11), had been presented by Frézier in 1737–9, although Danyzy's own paper was not finally printed until after the *Essai*. Thus, although Coulomb quotes neither Danyzy nor Frézier, it is very likely that he would have known of the collapse of arches by the formation of hinges.

In the summer of 1758, when he was just 22, Coulomb left Montpellier and returned to Paris, determined to join the *École du Génie*. If AUGOYAT is correct in stating that admission to Mézières after 1756 could only be obtained from the school at La Fère, then Coulomb must have gone there, although Gillmor states that candidates could be tutored in Paris for the examination set by Camus. In any case, Coulomb passed this examination a year after he left Montpellier, and entered the school at Mézières in February 1760, when he was nearly 24. He graduated almost two years later, in November 1761, with a monetary prize and an indifferent report from Chastillon.

Although Bossut may not have yet had much effect in modifying the strongly practical syllabus at Mézières, it may be guessed that his personal influence on Coulomb was strong. Bossut was only six years older than Coulomb, being under 30 when Coulomb started the courses. Bossut as a young man cannot have read widely, nor reflected deeply; he would have expounded the findings of La Hire and of Bélidor, and indeed interpreted them, but he may not have discovered the forty-year old work of Parent, and Coulomb could have left. Mézières unaware that the bending problem had been solved. Instead, he would have had on the one hand the essentially theoretical approach of Galileo, and on the other the practical formulae of Bélidor and of La Hire, together with some knowledge of the scientific contributions of Gregory, Bernoulli, Euler, and so on, in short, of the authorities whom he cites in the *Essai*.

In his first posting, to Brest, Coulomb would have needed to use little of the theory on the subjects which interested him in the *Essai* of 11 years later; he spent two years on coastal mapping before going to Martinique. On the other hand, his return from Martinique was in June 1772, only 9 months before he read the *Essai* to the *Académie*, and there is a question of when he found time to make his tests on the fracture of stone. Gillmor states that he did make experiments of some sort during the nine months, but he also makes it clear that it was Coulomb's habit in later postings to attempt to set up laboratory facilities, and to carry out experiments (e.g. on friction in 1779–80 while he was at Rochefort) on subjects unconnected with his direct military duties. Thus it is at least possible that the tests on Bordeaux stone (p. 6; p. 45) and Provence bricks (p. 7; p. 45) were made during Coulomb's first posting to Brest, 1762–3.

Coulomb's assignment to duty in Martinique was accidental. A ship was sailing for Martinique from Brest in February 1764, and the engineer who had been posted fell ill; Coulomb was drafted in his place. Major fortifications were proposed for Martinique to defend it against possible renewed attack by the English, and some difficulty was encountered in settling on a final design. A decision was finally taken in April 1764 in favour of a design which Coulomb supported, and he was put in charge of the work. He stayed on the island for nine years, employed on 'travaux pénibles sous un ciel brulant', to which he devoted himself, in the words of the *Éloge*: 'without reserve. That spirit of research, of experiment and theory, which so eminently distinguished him, he was able to apply only to the means of executing with greater economy and strength those works under his direction.'

The 'Essai'

As has been noted, Marshal Vauban had perfected that system of ditches, within which defenders could manoeuvre, and from which they could fire on attackers without themselves being exposed. These fortifications gave rise to some of the major problems of civil engineering; the walls of the ditches had to be retained, so that knowledge was required of 'la poussée des terres', and the construction of masonry arches was involved, so that knowledge was required of 'la poussée des voûtes'. Vauban had given his own rules, and Bélidor's *Science des Ingénieurs* gave adequate tables both for retaining walls and for masonry arches.

It is not too fanciful to suppose Coulomb in Martinique with a copy of Bélidor in his pocket, and to sense his dissatisfaction with these practical but efficient rules, both as a theorist with a strong bent to mathematics, and as a practical engineer concerned with economy and strength. There is every reason then for Coulomb having written the *Essai* for his own use (p. 4; p. 43) as an attempt to provide more rational solutions to the problems posed by the tasks in which he was engaged. It was these solutions that he presented to the *Académie*, as a *savan étranger*, on 10 March and 2 April 1773.

In accordance with its usual practice, the *Académie* appointed two referees to examine and report on Coulomb's *Essai*. These referees were de Borda, *pensionnaire* in the class of *géométrie*, and Bossut, *associé* in the same class. They reported very favourably, a year later, in April 1774, with a recommendation for publication. Further, by a rule of 1753, each *académicien* had the right to propose a name as *correspondant*, and such a proposal had to be approved by a two thirds majority of the whole body; Bossut proposed Coulomb as his own *correspondant*, and he was duly elected on 6 July 1774. In subsequent *séances* of the *Académie*, Coulomb would sit at a table directly behind Bossut, and could himself present *Mémoires* to the *Académie*.

The outstanding feature of the *Essai* is, of course, the use of limiting principles. No previous writer had allowed the plane of slip behind a retaining wall to enter the problem in terms of an arbitrary parameter, the actual plane being determined finally by use of variational methods to find a maximum (or minimum). As Coulomb notes in his own introduction, this technique is common to his attack on the problems of column fracture and of the collapse of arches. Coulomb uses these ideas with skill, but he does not begin to compete with the mathematical ability of Euler or of the Bernoullis; mathematically, the *Essai* is of negligible importance. However, whereas Euler had solved (for example) the general mathematical problem of the elastica, and had then coarsened the solution so that it could be applied to a model more or less representative of something real (the buckling of an elastic column), all Coulomb's problems in the *Essai* arose directly from engineering experience. He was not interested in 'applied mathematics', but in the use of mathematics to obtain solutions to actual practical problems.

Coulomb's later life

Gillmor gives a detailed biography after Coulomb's return to France, and only the main events need be rehearsed here. From 1773 to 1774 Coulomb was posted to Bouchain, and for the two following years to Cherbourg. While at Cherbourg he wrote his essay on magnetic compasses, which he submitted for the prize offered by the *Académie* in 1777. The essay shared the prize (with van Swinden) and was Coulomb's first excursion into physics.

In 1777 Coulomb was moved to Besançon, where he worked on several memoirs, including one on caissons for underwater work, published by the *Académie* in 1779. Altogether he read 6 *Mémoires* to the *Académie* between 1774 and 1781, although most of these were not published. In 1779 Coulomb started work at Rochefort on the experiments on friction for his *Théorie des machines simples*, an essay which won the *Académie* prize for 1781. These academic successes encouraged Coulomb to apply for permanent transfer to Paris, and this was granted in the autumn of 1781. At the same time, he was elected to the *Académie* as *adjoint mécanicien* (12 December 1781); his army rank was *Capitaine en Premier de la Première Classe* in the *Corps Royal du Génie*.

Coulomb was made *associé mécanicien* on 19 January 1784, and *associé de la classe de mécanique* in the reorganization of the *Académie* of 23 April 1785; in 1786 he was promoted to the rank of Major. Between 1785 and 1791 he presented the famous series of seven memoirs on electricity and magnetism. (Gillmor lists 32 memoirs read by Coulomb to the *Académie* between 1773 and 1806. Seven (including the *Essai*) were read before he was made a member in 1781, and sixteen between 1781 and the abolition of the *Académie* in 1793; nine were read to the Institut between 1795 and 1806. Of these thirty-two, twenty-two were in the field of physics, nine in mechanics, and one in plant physiology.)

Coulomb tendered his resignation from the army in 1790, giving as one reason that of ill-health. His resignation was accepted on 1 April 1791, and he was awarded an annual pension of 2240 *livres* in January 1792, after 31 years of service since he first entered the school at Mézières. (This pension was subsequently reduced.) In 1793, when the *Académie* shut its doors, Coulomb went with Borda to Coulomb's property at Blois, living well away from a dangerous Paris. It was here that he made experiments on the circulation of sap which led to the anomalous *Mémoire* on botany.

He returned to Paris in 1795, and was elected on 9 December a *membre de la section de physique expérimentale* of the *Institut*. He remained in Paris until his death in 1806.

Coulomb's health had been ruined in Martinique. Delambre says that he was attacked there by *maladies cruelles*, and that the members of the *Institut* had been anxious about him for a long time. Coulomb himself always thought that death was more or less close at hand; he died slowly and painfully at the age of seventy.

References

AMONTONS, G., De la résistance causée dans les machines, tant par les frottemens des parties qui les composent, que par la roideur des Cordes qu'on y employe, et la maniere de calculer l'un et l'autre, *Histoire de l'Académie Royale des Sciences, 1699*, 206, Paris (1702).

ANTONI, PAPACINO D', see PAPACINO

ARTZ, F. B., *The development of technical education in France 1500–1850*, The M.I.T. Press (1966).

AUGOYAT, A. M., *Aperçu historique sur les fortifications, les ingénieurs et sur le corps du génie en France* (3 vols.), Paris (1860–4).

BAKER, B., The actual lateral pressure of earthwork, *Min. Proc. Instn Civ. Engrs.* **65**, 140 (1880).

BAKER, J. F., HORNE, M. R. and HEYMAN, J., *The steel skeleton*, vol. 2: *Plastic behaviour and design*, Cambridge (1956).

BALBO, P., Vita di Alessandro Vittorio Papacino d'Antoni, *Mémoires de l'Académie Impériale des Sciences, Littérature et Beaux-arts de Turin, pour les années XII et XIII* (classe de littérature et beaux-arts), 281–376, Turin (*An* XIII-1805).

BARLOW, P., *An essay on the strength and stress of timber*, London (1817).

BAXTER, J. W., GEE, A. F. and JAMES, H. B., Gladesville Bridge, *Proc. Instn Civ. Engrs.* **30**, 489 (1965).

BELGRADO, J., *De corporibus Elasticis Disquisito Physico-mathematica*, Parma (1748).

BÉLIDOR, B. F. DE, *La science des ingénieurs dans la conduite des travaux de fortification et d'architecture civile*, Paris (1729).

BÉLIDOR, B. F. de, *Dictionnaire portatif de l'ingénieur*, Paris (1755).

BERNOULLI, DANIEL, De vibrationibus et sono laminarum elasticarum, *Commentarii Academiae Scientiarum Imperialis Petropolitanae, 1741–3*, **13**, 105–20, Petersburg (1751).

BERNOULLI, JAMES, Specimen alterum calculi differentialis (*Acta eruditorum Lipsiae 1691*); in *Opera* (2 vols.), no. XLII, vol. 1, p. 442, Geneva (1744).

BERNOULLI, JAMES, Curvatura laminae elasticae (*Acta eruditorum Lipsiae 1694*); in *Opera* (2 vols.), no. LVIII, vol. 1, p. 576, Geneva (1744).

BERNOULLI, JAMES, Explicationes, annotationes et additiones (*Acta eruditorum Lipsiae 1695*); in *Opera* (2 vols.), no. LXVI, vol. 1, p. 639, Geneva (1744).

BERNOULLI, JAMES, Véritable Hypothèse de la Résistance des Solides, Avec la Démonstration de la Courbure des Corps qui font ressort (*Histoire de l'Académie des Stiences, Paris, 1705*); in *Opera* (2 vols.), no. CII, vol. 2, p. 976, Geneva (1744).

BERNOULLI, JAMES, Problema de Curvatura fornicis, cujus partes se mutuo proprio pondere suffulciunt sine opere caementi; in *Opera* (2 vols.) no. CIII, Varia Posthuma, vol. 2, Articul. 29, p. 1119, Geneva (1744).

BERTOT, H., *Mémoires et Compte Rendu des travaux de la Société des Ingénieurs Civils*, pp. 277–80, Paris (1855).

BIOT, J.-B., *Mélanges scientifiques et littéraires* (3 vols.), Paris (1858).

BLONDEL, F., Resolution des quatre principaux problemes d'architecture, *Mémoires de l'Académie Royale des Sciences, 1666–99*, **5**, Paris (1729).

References

BOISTARD, L. C., Expériences sur la stabilité des voûtes, *see* LESAGE, **2**, 171.

BOSSUT, C., Recherches sur l'équilibre des voûtes, *Histoire de l'Académie Royale des Sciences, 1774*, 534, Paris (1778).

BOSSUT, C., Nouvelles recherches sur l'équilibre des voûtes en dôme, *Histoire de l'Académie Royale des Sciences, 1776*, 587, Paris (1779).

BOSSUT, C. and VIALLET, G. *Recherches sur la construction la plus avantageuse des digues*, Paris (1764).

BOUASSE, M. H., Sur la théorie des déformations permanentes de Coulomb. Son application à la traction, la torsion et le passage à la filière, *Annales de Chimie et de Physique*, 7th series, **23**, 198–240 (1901).

BOUSSINESQ, J., *Essai théorique sur l'équilibre d'élasticité des massifs pulvérulents et sur la poussée des terres sans cohésion*, Paris (1876).

BOUSSINESQ, J., Note on Mr G. H. Darwin's paper 'On the horizontal thrust of a mass of sand', *Minutes of Proceedings Inst. Civ. Engrs.* **72**, 262 (1882–3).

BOUSSINESQ, J., Sur la détermination de l'épaisseur minimum que doit avoir un mur vertical, d'une hauteur et d'une densité données, pour contenir un massif terreux, sans cohésion, dont la surface supérieure est horizontale, *Annales des Ponts et Chaussées*, 6th series, **3**, 625 (1882).

BOUSSINESQ, J., Note sur la poussée horizontale d'une masse de sable à propos des expériences de M. Darwin; Addition relative aux expériences de M. Gobin, *Annales des Ponts et Chaussées*, 6th series, **6**, 494, 510 (1883).

BOUSSINESQ, J., Complément à de précédentes notes sur la poussée des terres, *Annales des Ponts et Chaussées*, 6th series, **7**, 443 (1884).

BUFFON, G.-L. L., Expériences sur la force du bois, *Histoire de l'Académie Royale des Sciences, 1740*, 543, Paris (1742). Second Mémoire, *ibid., 1741*, 292, Paris (1744).

BÜLFFINGER, G. B., De solidorum resistentia specimen, *Commentarii Academiae Scientiarum Imperialis Petropolitanae, 1729*, **4**, Petersburg (1735).

BULLET, P., *L'architecture pratique*, Paris (1691).

CALLADINE, C. R., A microstructural view of the mechanical properties of saturated clay, *Géotechnique*, **21**, 391 (1971).

CAQUOT, A., *Équilibre des massifs à frottement interne. Stabilité des terres pulvérulentes et cohérentes*, Paris (1934).

CAUCHY, A. L., Recherches sur l'équilibre et le mouvement intérieur des corps solides ou fluides, élastiques ou non élastiques, *Bull. Soc. Philomath. Paris*, 9–13 (1823). See also MOIGNO.

CHEN, W. F., GIGER, M. W. and FANG, H. Y., On the limit analysis of stability of slopes, *Soils and Foundations, The Japanese Society of Soil Mechanics and Foundation Engineering*, **9**, 23–32 (1969).

CLAPEYRON, B. P. E., Calcul d'une poutre élastique reposant librement sur des appuis inégalement espacés, *Comptes Rendus hebdomadaires des Séances de l'Académie des Sciences*, **45**, 1076–80, Paris (1857).

CLAUSEN, T., Ueber die Form architektonischer Säulen, *Bulletin de la classe Physico-mathématique de l'Académie Impériale des Sciences de Saint-Pétersbourg*, **9**, 370 (1851).

COLLIN, A., *Recherches expérimentales sur les glissements spontanés des terrains argileux...*, Paris (1846). (Translated by W. R. Schriever, *Landslides in clays*, Toronto (1956).)

CONSIDÈRE, A., Sur la poussée des terres, *Annales des Ponts et Chaussées*, 4th series, **19**, 547 (1870).

COULOMB, C. A., Théorie des machines simples en ayant égard au frottement de leurs parties et à la roideur des cordages, *Mémoires de Mathématique et de Physique*,

References

présentés à l'Académie Royale des Sciences par divers Savans, et lus dans ses assemblées, **10**, 161–332, Paris (1785).

COULOMB, C. A., *Théorie des machines simples...*, Paris (1821).

COUPLET, P., De la poussée des terres contre leurs revestemens, et de la force des revestemens qu'on leur doit opposer, *Histoire de l'Académie Royale des Sciences, 1726*, 106, Paris (1728); *1727*, 139, Paris (1729); *1728*, 113, Paris (1730).

COUPLET, P., De la poussée des voûtes, *Histoire de l'Académie Royale des Sciences, 1729*, 79, Paris (1731), and *1730*, 117, Paris (1732).

CULMANN, K., *Die graphische Statik*, Zürich (1866).

DANYZY, A.-A.-H., Méthode générale pour déterminer la résistance qu'il faut opposer à la poussée des voûtes, 27 Feb. 1732, *Histoire de la Société Royale des Sciences établie à Montpellier*, **2**, 40, Lyon (1778).

DARWIN, G. H., On the horizontal thrust of a mass of sand, *Minutes of Proceedings Inst. Civ. Engrs.* **71**, 350 (1882–3).

DE JOSSELIN DE JONG, G., *Statics and kinematics in the failable zone of a granular material*, Delft (1959).

DELAMBRE, J.-B., Éloge historique de M. Coulomb, *Histoire de la classe des Sciences Mathématiques et Physiques de l'Institut National de France*, **7**, 206, Paris (1806).

DESAGULIERS, J. T., *A treatise of the motion of water...by E. Mariotte...*, London (1718).

DRUCKER, D. C., Limit analysis of two and three dimensional soil mechanics problems, *Journal of the Mechanics and Physics of Solids*, **1**, 217–26 (1953).

DRUCKER, D. C., Coulomb friction, plasticity, and limit loads, *J. appl. Mech.* **21**, 71–4 (1954).

DRUCKER, D. C. and PRAGER, W., Soil mechanics and plastic analysis or limit design, *Quart. Appl. Math.* **10**, 157–65 (1952).

ENCYCLOPAEDIA BRITANNICA, Supplement to the 4th, 5th and 6th editions, article 'Coulomb', vol. 3, pp. 414–19 (1824).

EULER, L., Solutio problematis de invenienda curva..., *Commentarii Academiae Scientiarum Imperialis Petropolitanae, 1728*, **3**, 70, Petersburg (1732).

EULER, L., *Methodus inveniendi lineas curvas Maximi Minimive proprietate gaudentes, sive solutio problematis isoperimetrici latissimo sensu accepti*, Lausanne and Geneva (1744).

EULER, L., Sur la force des Colonnes, *Mémoires de l'Académie Royale des Sciences et Belles Lettres*, **13**, 252, Berlin (1757).

EULER, L., De pressione funium tensorum in corpora subjecta,... *Novi Commentarii Academiae Scientiarum Imperialis Petropolitanae*, **20**, 304, 1775, Petersburg (1776).

EULER, L., (translated by) OLDFATHER, W. A., ELLIS, C. A. and BROWN, D. M., Leonhard Euler's elastic curves, *Isis*, **20**, 72–160 (1933).

EWING, J. A., *The strength of materials*, Cambridge (1899).

FELD, J., Early history and bibliography of soil mechanics. *Proc. 2nd Int. Conf. Soil Mechanics*, p. 1, Rotterdam (1948).

FELLENIUS, W., *Erdstatische Berechnung...*, Berlin (1927).

FITCHEN, J., *The construction of Gothic cathedrals*, Oxford (1961).

FRANKL, P., *The Gothic, literary sources and interpretations through eight centuries*, Princeton (1960).

FRÉZIER, A. F., *La théorie et la pratique de la coupe des pierres et des bois pour la construction des voûtes et autres parties des batimens civils et militaires, ou traité de stereotomie à l'usage de l'architecture* (3 vols.), Strasbourg and Paris (1737–9).

GALILEO, *Discorsi e Dimostrazioni matematiche, intorno à due nuove scienze attenenti alla mecanica e i movimenti locali*, Leida (1638).

GALILEO, *Dialogues concerning two new sciences*, translated by H. Crew and A. de Salvio, New York (1952).

References

GAUTHEY, E.-M., *Traité de la construction des ponts* (edited by C. L. M. H. Navier), (2 vols.), Paris (1809, 1813).

GAUTIER, H., *Dissertation sur l'epaisseur des culées des ponts,*... Paris (1717).

GILLMOR, C. S. Charles Augustin Coulomb: Physics and engineering in eighteenth century France, Dissertation for degree of Ph.D., Princeton University (1968); University Microfilm No. 69–2741.

GILLMOR, C. S., *Coulomb and the evolution of physics and engineering in eighteenth-century France*, Princeton University Press (1971).

GIRARD, P. S., *Traité analytique de la résistance des solides, et des solides d'égale résistance*, Paris (1798).

GREGORY, D., Catenaria, *Phil. Trans.* no. 231, p. 637 (1697).

HAMILTON, S. B., Charles Auguste de Coulomb: A bicentenary appreciation of a pioneer in the science of construction, *Trans. Newcomen Soc.* **17**, 27 (1936–7).

HAMILTON, S. B., The French civil engineers of the eighteenth century, *Trans. Newcomen Soc.* **22**, 149 (1941–2).

HANN, J., *Bridges. Section I–Theory* (Theory of bridges, by James Hann; general principles of construction, etc., translated from Gauthey; theory of the arch, etc., by Professor Moseley; papers on foundations, by T. Hughes; etc.), London (John Weale) (1843).

HEATHER, J. F., *An elementary course of mathematics, prepared for the use of the Royal Military Academy*, vol. 3, part 1, *Mechanics*, London (John Weale) (1853).

HEYMAN, J., The stone skeleton, *Int. J. Solids Structures*, **2**, 249 (1966).

HEYMAN, J., On shell solutions for masonry domes, *Int. J. Solids Structures*, **3**, 227 (1967).

HEYMAN, J., Westminster Hall roof, *Proc. Instn Civ. Engrs.* **37**, 137 (1967).

HEYMAN, J., On the rubber vaults of the Middle Ages, and other matters, *Gaz. Beaux-Arts*, **71**, 177, March (1968).

HEYMAN, J., The safety of masonry arches, *Int. J. mech. Sci.* **11**, 363 (1969).

HEYMAN, J., *Plastic design of frames*, vol. 2, *Applications*, Cambridge (1971).

HODGKINSON, E., On the transverse strain, and strength of materials, *Memoirs of the Literary and Philosophical Society of Manchester*, 2nd series, **4**, 225 (1824).

HODGKINSON, E., Theoretical and experimental researches to ascertain the strength and best forms of iron beams, *Memoirs of the Literary and Philosophical Society of Manchester*, 2nd series, **5**, 407 (1831).

HOLLISTER, S. C., The Life and Works of Charles Augustin Coulomb, *Mechanical Engineering*, 615 (1936).

HOOKE, R., *A description of helioscopes, and some other instruments*, London (1676) (*sic*, actually 1675). See GUNTHER, R. T., *Early science in Oxford*, vol. 8, pp. 119–52 (1931).

HUYGENS, C., *Oeuvres complètes*, vol. 16, La Haye (1929); vol. 19, La Haye (1937).

JENKIN, H. C. F., Article 'Bridges', *Encyclopaedia Britannica*, 9th edition, Edinburgh (1876).

KÁRMÁN, TH. VON, Festigkeitsversuche unter allseitigem Druck, *Zeitschrift des vereines Deutscher Ingenieure*, **55**, no. 42, 1749, October (1911).

KELLER, J., The shape of the strongest column, *Arch. Rational Mech. Anal.* **5**, 275–85 (1960).

KERISEL, J., Historique de la mécanique des sols en France jusqu'au 20e siècle, *Géotechnique*, **6**, 151 (1956).

KOOHARIAN, A., Limit analysis of voussoir (segmental) and concrete arches, *Proc. Am. Concr. Inst.* **89**, 317 (1953).

LAGRANGE, J. L., Sur la figure des colonnes, *Miscellanea Taurinensia*, V, 123, Turin (1770–73).

References

LA HIRE, P. DE, *Traité de Mécanique*, Paris (1695).

LA HIRE, P. DE, Sur la construction des voûtes dans les édifices, *Mémoires de l'Académie Royale des Sciences, 1712*, 69, Paris (1731).

LAMÉ, M. G. and CLAPEYRON, E., Mémoire sur la stabilité des voûtes, *Annales des Mines*, **8**, 789 (1823).

LEIBNITZ, G. W., Demonstrationes novae de Resistantiâ solidorum, *Acta Euriditorum Lipsiae*, 319, July (1684).

LESAGE, P. C., *Recueil de divers mémoires extraits de la Bibliothèque Impériale des Ponts et Chaussées à l'usage de MM. les ingénieurs* (2 vols.), Paris (1810).

LE SEUR, T., JACQUIER, F. and BOSCOVICH, R. G., *Riflessioni...sopra alcune difficoltà spettanti i danni, e risarcimenti della cupola di S. Pietro*, Rome (1743).

LE SEUR, T., JACQUIER, F. and BOSCOVICH, R. G., *Parere di tre mattematici sopra i danni che si sono trovati nella cupola di S. Pietro sul fine dell'Anno 1742*, Rome (1743).

LÉVY, M., Sur une théorie rationnelle de l'équilibre des terres fraîchement remuées et ses applications au calcul de la stalibité des murs de soutènement, *Journal de Mathématiques*, 2nd series, **18**, 241 (1873). (See also *Comptes Rendus, Acad. Sci.* **68**, 1456 (1869).)

MAINDRON, E., *L'Académie des Sciences, Histoire de l'Académie...*, Paris (1888).

MAINDRON, E., *L'ancienne Académie des Sciences, Les Académiciens 1666–1793*, Paris (1895).

MARCHETTI, A., *De resistentia solidorum*, Florence (1669).

MARIOTTE, E., *Traité du mouvement des eaux*, Paris (1686). See also DESAGULIERS.

MAYNIEL, K., *Traité expérimental, analytique et pratique de la poussée des terres et des murs de revêtement,...*, Paris (1808).

MOHR, O., Über die Darstellung des Spannungszustandes und des Deformationszustandes eines Körperelementes und über die Anwendung derselben in der Festigkeitslehre, *Der Civilingenieur*, **28**, col. 113–56 (1882).

MOIGNO, F. N. M., *Leçons de mécanique analytique, rédigées principalement d'après les méthodes d'Augustin Cauchy,...*, (Paris 1868).

MOSELEY, H., *The mechanical principles of engineering and architecture*, London (1843).

MUSSCHENBROEK, P. VAN, *Physicae experimentales, et geometricae,... Dissertationes*, Leyden (1729).

MUSSCHENBROEK, P. VAN, *Essai de physique*, Leyden (1739).

NAVIER, C. L. M. H., Rapport fait a l'Académie des Sciences le lundi 4 septembre 1820, sur un Mémoire de M. Navier qui traite de la Flexion des lames élastiques, *Annales de Chimie et de Physique*, **15**, 264, Paris (1820).

NAVIER, C. L. M. H., Note sur les questions de statique dans lesquelles on considère un corps pesant supporté par un nombre de points d'appui surpassant 3, *Nouveau Bulletin des Sciences par la Société Philomatique de Paris*, 35–7 (1825).

NAVIER, C. L. M. H., Sur la flexion des verges élastiques courbes (Extrait d'un Mémoire présenté à l'Académie des Sciences, le 23 novembre 1819), *Nouveau Bulletin des Sciences par la Société Philomatique de Paris*, 98–100, 114–18 (1825).

NAVIER, C. L. M. H., *Résumé des leçons données à l'École des Ponts et Chaussées, sur l'application de la mécanique à l'établissement des constructions et des machines*, 2nd edition, Paris (1833).

NAVIER, C. L. M. H., *Résumé des leçons etc.*, 3rd edition, with notes and appendices by Saint-Venant, Paris (1864).

ORAVAS, G. AE. and MCLEAN, L., Historical development of energetical principles in elastomechanics, *Applied Mechanics Reviews*, **19**, no. 8, 647–58, no. 11, 919–33 (1966).

PALMER, A. C., Private communication.

References

PAPACINO D'ANTONI, A. V., *Instituzioni fisico-meccaniche per le Regie Scuole d'Arti-glieria, e Fortificazione* (2 vols.), Torino (1773, 1774).

PAPACINO D'ANTONI, A. V., *Dell'Architettura militare* (6 vols.), Torino (1778, 1779, 1759 [*sic*], 1780, 1781, 1782).

PARENT, A., *Essais et recherches de Mathematique et de Physique* (3 vols.), Paris (1713).

PERRONET, J. R., Mémoire sur la réduction de l'épaisseur des piles, et sur la courbure qu'il convient de donner aux voûtes, pour que l'eau puisse passer plus librement sous les ponts, *Histoire de l'Académie Royale des Sciences, 1777*, 553, Paris (1780).

PETTERSON, K. E., The early history of circular sliding surfaces, *Géotechnique*, **5**, 275 (1955).

PIPPARD, A. J. S., TRANTER, E. and CHITTY, L., The mechanics of the voussoir arch, *J. Instn Civ. Engrs.* **4**, 281 (1936).

PIPPARD, A. J. S. and ASHBY, R. J., An experimental study of the voussoir arch, *J. Instn Civ. Engrs.* **10**, 383 (1938).

POLENI, G., *Memorie istoriche della gran cupola del Tempio Vaticano*, Padova (1748).

PONCELET, J. V., *Mécanique Industrielle*, Liége (1839).

POTIER, A. [Preface to the] *Collection de Mémoires rélatifs à la Physique...* vol. I, *Mémoires de Coulomb*, Paris (1884).

PRAGER, W., *An introduction to plasticity*, Addison–Wesley (1959).

PRONY, R. DE, *Mécanique philosophique*, Paris (*An* VIII-1800).

PRONY, R. DE, *Recherches sur la poussée des terres, et sur la forme et les dimensions à donner aux murs de revêtement*, Paris (*An* X-1802).

RANKINE, W. J. M., On the stability of loose earth, *Phil. Trans. Roy. Soc.* **147**, 9 (1857).

RONDELET, J., *Traité théorique et pratique de l'art de bâtir*, 5th edition (5 vols. plus plates), Paris (1812–14).

ROSCOE, K. H., SCHOFIELD, A. N. and WROTH, C. P., On the yielding of soils, *Géotechnique*, **8**, 22 (1958).

SAINT-VENANT, B. DE, Rapport fait à l'Académie des Sciences sur un Mémoire de M. Maurice Lévy... par MM. Combes, Serret, Bonnet, Phillips, de Saint-Venant rapporteur, *Journal de Mathématiques*, 2nd series, **15**, 237 (1870).

SAINT-VENANT, B. DE, Sur une détermination rationnelle, par approximation, de la poussée qu'exercent des terres dépourvues de cohésion, contre un mur ayant une inclinaison quelconque, *Journal de Mathématiques*, 2nd series, **15**, 250 (1870). (See also *Comptes Rendus, Acad. Sci.*, **70**, 229–35, 281–6 (1870).)

SAINT-VENANT, B. DE, see NAVIER, *Résumé des leçons etc.*

SCHOFIELD, A. N. and WROTH, C. P., *Critical state soil mechanics*, London (1968).

SÉJOURNÉ, P., *Grandes voûtes* (6 vols.), Bourges (1913–16).

SKEMPTON, A. W., Alexandre Collin, 1808–90, Pioneer in soil mechanics, *Trans. Newcomen Soc.* **25**, 91, 1945–6 and 1946–7, London (1950).

SKEMPTON, A. W., The $\phi = 0$ analysis of stability and its theoretical basis, *Proc. 2nd Int. Conf. Soil Mechanics*, p. 72, Rotterdam (1948).

SOKOLOVSKII, V. V., *Statics of soil media* (translated by D. H. Jones and A. N. Schofield), London (1960).

STIRLING, J., *Lineae Tortii Ordinis Neutonianae*, Oxford (1717).

STRAUB, H., *A History of civil engineering*, London (1952). [*Die Geschichte der Bauingenieurkunst*, Basle (1949).]

TAYLOR, D. W., *Fundamentals of soil mechanics*, New York (1948).

TERZAGHI, K., *Theoretical soil mechanics*, New York (1943).

TIMOSHENKO, S. P., *History of strength of materials*, McGraw-Hill (1953).

References

TODHUNTER, I. and PEARSON K., *A history of the theory of elasticity, and of the strength of materials, from Galilei to the present time* (3 vols.), Cambridge (1886–93).

TREDGOLD, T., *A practical essay on the strength of cast iron*, London (1822).

TRUESDELL, C., *The rational mechanics of flexible or elastic bodies 1638–1788*, Introduction to Leonhardi Euleri Opera Omnia, 2nd series, vol. 11 (2), Zürich (1960).

VARIGNON, P., De la résistance des solides..., *Histoire de l'Académie Royale des Sciences, 1702*, Paris (1704).

VAUBAN, MARQUIS DE, *Traité de l'attaque des places*, Paris (1704); *Traité de la défense des places*, Paris (1706).

WARE, S., *A treatise of the properties of arches, and their abutment piers*, London (1809).

YOUNG, T., *Miscellaneous works* (3 vols.), London (1855).

YVON VILLARCEAU, A., L'établissement des arches de pont, *Institut de France, Académie des Sciences, Mémoires présentés par divers savants*, **12**, 503 (1854).

Name index

Amontons, G., 5, 44, 75–6, 199
Antoni, Papacino d', *see* Papacino
Artz, F. B., viii, 199
Ashby, R. J., 188, 204
Augoyat, A. M., viii, 194, 199

Baker, B., 156, 188, 199
Baker, Sir John, 164, 199
Balbo, P., 129, 199
Barlow, P., 100, 102, 199
Baxter, J. W., 189, 199
Belgrado, J., 89, 199
Bélidor, B. F. de, 39, 69, 75, 77, 84ff,
    97, 99, 126, 170, 172, 174, 181, 186,
    192, 195–6, 199
Bernoulli, Daniel, 81, 105ff, 196,
    199
Bernoulli, James, 31, 63, 73, 75, 81–2,
    94, 96, 99, 105–6, 182, 195–6, 199
Bertot, H., 109, 199
Biot, J.-B., 190, 199
Blondel, F., 86, 91, 99, 124, 199
Boistard, L. C., 183, 185–6, 188, 200
Borda, J. C. de, 196–7
Boscovich, R. G., 175, 203
Bossut, C., viii, 10, 47, 75, 77ff, 89, 132,
    161, 181, 192, 194, 196, 200
Bouasse, M. H., 122, 200
Boussinesq, J., vii, 149, 152ff, 158, 200
Brown, D. M., 201
Brunel, I. K., 188
Buffon, G. L. L., 97, 99, 200
Bülffinger, G. B., 96–7, 101, 200
Bullet, P., vii, 70, 124, 200

Calladine, C. R., 161, 200
Camus, C. E. L., 191–2, 194
Caquot, A., 155, 200
Cauchy, A. L., 71, 112, 200
Chastillon, N. de, 192, 194
Chen, W. F., 141, 143, 200
Chitty, L., 188, 204
Clapeyron, B. P. E., 109, 175, 185–6,
    200, 203

Clausen, T., 109, 200
Collin, A., 155, 200
Considère, A., 149, 151–2, 200
Couplet, P., viii, 126ff, 162, 170ff, 175,
    176, 181–6, 201
Crew, H., 201
Culmann, K., 113, 201

d'Alembert, J. le R., 193
Danyzy, A.-A.-H., 175, 184, 194, 201
Darwin, G. H., 155, 201
de Josselin de Jong, G., 156, 201
Delambre, J. B., 190–1, 201
Desaguliers, J. T., 201, 203
de Salvio, H., 201
Drucker, D. C., 73, 140, 142, 147–8,
    151, 164, 201
Dupin, C., 186

Ellis, C. A., 201
Euler, L., 30, 62, 75, 80, 81, 99, 105ff,
    169, 195–6, 201
Ewing, J. A., 188, 201

Fang, H. Y., 141, 143, 200
Feld, J., 201
Fellenius, W., 156, 201
Fitchen, J., 183, 201
Frankl, P., 189, 201
Frézier, A. F., 175, 194, 201

Gadroy, —, 131, 148
Galileo, viii, 80, 82, 87, 90–3, 98–9,
    102, 104, 189, 195, 201
Gauthey, E.-M., 148, 184, 186–7,
    202
Gautier, H., 124, 169, 202
Gee, A. F., 189, 199
Giger, M. W., 141, 143, 200
Gillmor, C. S., viii, 71, 190ff, 202
Girard, P. S., viii, 98ff, 202
Gregory, D., 3, 42, 70, 75–6, 81, 165,
    168, 178, 195, 202
Gunther, R. T., 202

Hamilton, S. B., 89, 190, 202
Hann, J., 187, 202
Harrison, T., 188
Heather, J. F., 187, 202
Heyman, J., 80, 163–4, 178, 189, 199, 202
Hodgkinson, E., 100, 102, 202
Hollister, S. C., 89, 190, 202
Hooke, R., 70, 76, 168, 202
Horne, M. R., 164, 199
Huygens, C., 91, 202

Jacquier, F., 175, 203
James, H. B., 189, 199
Jenkin, H. C. F., 188, 202
Jones, D. H., 204

Kármán, Th. von, 122, 202
Keller, J., 109, 202
Kerisel, J., 70–1, 155, 202
Kooharian, A., 164, 202

Lagrange, J. L., 99, 109, 192, 202
La Hire, P. de, vii, 39, 40, 69, 74–6, 82–6, 94, 168–70, 174, 176, 181–2, 184, 186, 189, 194, 203
Lamé, M. G., 175, 185–6, 203
Laplace, P. S., 192
Leibnitz, G. W., 93–4, 99, 203
Le Monnier, P. C., 194
Lesage, P. C., 183, 203
Le Seur, T., 175, 203
Lévy, M., 149ff, 203

McLean, L., 91, 203
Maindron, E., viii, 193, 203
Marchetti, A., 91, 99, 203
Mariotte, E., 91ff, 98–9, 102–3, 192, 203
Maxwell, James Clerk, 155
Mayniel, K., viii, 124, 130–1, 148–9, 203
Mohr, O., 113, 203
Moigno, F. N. M., 200, 203
Monge, G., 192
Moseley, H., 86, 187, 203
Musschenbroek, P. van, 13, 50, 71, 75, 80–1, 97, 99, 101, 110, 203

Navier, C. L. M. H., 89, 101ff, 109, 184, 186–7, 202, 203

Oldfather, W. A., 201
Oravas, G. Ae., 91, 203

Palmer, A. C., 143, 203
Papacino d'Antoni, A. V., 129–30, 148, 204
Parent, A., 87, 89, 92, 94, 96ff, 101, 103, 169, 176, 195, 204
Pearson, K., 89, 90, 105, 205
Perronet, J. R., 183–4, 189, 204
Petterson, K. E., 156, 204
Pippard, A. J. S., 188, 204
Poleni, G., viii, 76, 97, 175ff, 181, 186, 204
Poncelet, J. V., 149, 187, 204
Potier, A., 190, 204
Prager, W., 116, 142, 164, 201, 204
Prony, R. de, 99, 111–12, 138, 148, 183, 192, 204

Querlonde, —, 130

Rankine, W. J. M., 136, 149, 204
Rennie, John, 188
Rochon, A. M. de, 193
Rondelet, J., 132, 148, 184, 204
Roscoe, K. H., 161, 204

Saint-Venant, B. de, viii, 89, 91, 92, 101ff, 109, 150–1, 186, 204
Sallonnyer, —, 130
Schofield, A. N., vii, viii, 156, 161, 204
Schriever, W. R., 200
Séjourné, P., 187, 204
Skempton, A. W., 156, 204
Sokolovskii, V. V., vii, 116, 145, 155ff, 204
Stephenson, R., 188
Stirling, J., 176–7, 204
Straub, H., 89, 204

Taylor, D. W., 141, 204
Telford, T., 188
Terzaghi, K., viii, 204
Timoshenko, S. P., 89, 190, 204
Todhunter, I., 89, 90, 105, 205
Tranter, E., 188, 204
tre mattematici, 175, 178, 186, 203
Tredgold, T., 100, 205
Truesdell, C., viii, 76, 168, 205

van Swinden, J. H., 197
Varignon, P., 94, 96, 99, 101–2, 192, 205
Vauban, Marquis de, 20, 54–5, 71, 75, 85, 124, 195–6, 205
Viallet, G., 75, 200

Ware, S., 75–6, 205
Weale, J., 187
Wroth, C. P., 156, 161, 204

Young, T., 109, 205
Yvon Villarceau, A., 165, 184, 205

Subject index

(Place names are included in this index)

abutment
  spread, 166–7
  thrust, 40, 69, 76, 82, 162ff, 185
Académie, organization of, 193
Alessandria, soil tests at, 148
Angoulême, 194
arches, vii, viii, 3, 28ff, 42–3, 61ff, 82,
  162ff, 194
  hinges in, 83, 163ff, 168, 172, 181,
    185, 194
  model tests on, 169, 175, 183–4, 188
  of minimum thickness, 168–9, 172–3
  rupture of, 4, 35ff, 39, 43, 66ff, 69,
    83
  stability of, 98, 163, 186
  thrust lines for, 164, 174, 181, 186, 188

beam tests, 87, 92–3, 97–8, 100
bending of beams, vii, viii, 82, 89ff,
  100ff
  fracture of cantilever, 7, 8ff, 45, 46–7
  of equal resistance, 90, 94, 124
Besançon, 197
Blois, 190, 198
Bordeaux, 6n, 45n, 110, 195
Bouchain, 197
bounds
  lower, 116, 132, 141
  upper, 70, 116–17, 132, 140–1
Brest, 195
brick, 5, 7, 13, 44–5, 50, 71, 80, 110,
  120–1, 195
buckling of strut, 80, 107ff, 125, 196

caissons, 197
cantilever beam, 7, 8ff, 45, 46–7, 81
  deflexion of, 108
catenary, 3, 30, 42, 62, 75–6, 81, 165,
  168, 177–8
centering, forces on, 183
characteristics, 72–3, 116, 156–7
Cherbourg, 197
Chester (Grosvenor Bridge), 188

cohesion, vii, 6–7, 44–5
  of brick, 7, 45, 71
  of mortar, 7, 45
  of stone in shear, 6, 45
  of stone in tension, 6, 45
continuum mechanics, 71, 112
Corps du Génie, 191, 197
Critical state theory, viii, 160–1

deflexion of beams, 108–9
dilatation, 119, 133, 146, 161
discontinuities in stress fields, 142, 153
dome
  analysis of, 175, 178, 182, 186
  of minimum thickness, 179ff
  thrust lines for, 178
dykes, 10, 19, 47, 54, 77ff, 132

École
  d'Artillerie, 192
  du Corps Royale du Génie, 191–2, 194
  des Mines, 192
  Polytechnique, 99, 192
  des Ponts et Chaussées, 99, 183–4, 192
elastic design, 102
elastica, 105ff, 196
elasticity, absolute, 106, 108
equilibrium
  geometry of, 71, 163
  in beam, 96, 97, 100
Euler buckling load, 107–8, 125

fan, 158
Fontainebleau, 191
foundations, 125
fracture
  angle, 11, 13, 48–9
  plane, 1, 41, 70, 133
  of cantilever, 7, 8ff, 45, 46–7, 49, 96, 98
  of pier, vii, 1, 10ff, 13, 41, 48ff, 50,
    98, 110ff
  'elastic', 9, 10, 46–7, 77, 89
  'rigid', 9, 10, 47, 77, 89

210

friction, vii, 5, 44, 76–7
  of bricks, 5, 44
  of ropes, 169
  of stone, 6, 44
  and plasticity theorems, 147–8
  on retaining wall, 20–1, 23, 54–5, 56,
    128–9, 144ff
friction circles, 112
funicular polygon, 82, 168

geometrical factors of safety, 189
Gladesville Bridge, 189

hinges,
  in arches, 83, 163ff, 168, 172, 181,
    185, 194
  in dome, 175
  plastic, 104, 188

ice, effect of, 23, 57
Intendant des Eaux..., 190–1
Iron Bridge (Coalbrookdale), 188

Jülich, soil tests at, 148–9

la Fère, 87, 192, 194
landslides, 155
ligne, 71
limit theorems, vii, viii, 70, 116, 147–8,
    160, 163–4, 168, 178, 189
logarithmic spiral, 143, 158
London Bridge, 188

Martinique, 7, 45, 110, 191, 195–6,
    198
masonic lodges, 189
masonry,
  low stress levels in, 74, 80, 163
  strength of, 98
material tests, 80, 91, 97, 110, 120
maximum shear stress, 113–14,
    141–2
mechanism (of collapse), 167–8, 172,
    175
Metz, 192
Mézières, 77, 191ff
Mises criterion, 118
model tests on arches, 169, 175
Mohr–Coulomb criterion, 113ff, 146,
    160, 164
Mohr's circle, 113ff, 136ff

Montpellier, 175, 194

natural slope of soils, 19, 22, 23, 54, 56,
    57, 85, 125–6
neutral axis, 87, 89, 92, 94–5, 98–9, 105
normality condition, 117–18

oscillation of beams, 108
overburden, 159
overturning moment (on retaining wall)
  Couplet, 128
  Bélidor, 129
  Papacino, 129–30
  Sallonnyer, 130–1
  Coulomb–Rankine, 135
  Coulomb, 138
  Boussinesq, 154

packing,
  pyramidal, 128
  tetrahedral, 127–8
pied, 71
pier, vii, 1, 10ff, 41, 48ff, 70
piles, 125
plane sections, 71, 101
plastic analysis of beams, 91–2
plastic modulus, 104–5
plasticity, limit theorems of, vii, viii,
    70, 116, 147–8, 160, 163–4, 168,
    178, 189
plate-bande, 3, 33, 43, 65, 84, 167,
    169, 175, 184
Pont du Gard, 183
Pont de Nemours, 184
Pont de Neuilly, 183
pore-water, 23, 57, 79, 161
potential function, 150
pouce, 71
principal
  axes, 102
  directions, 149
  stress difference, 113
  stresses, 113

Quebec oak, fracture strength, 101

Rankine states, 70, 72, 136ff, 147
retaining walls, 3, 15ff, 42, 51ff, 85,
    124ff
Rochefort, 195, 197
Royal Military Academy, 187

rupture
  plane, 2, 17, 42, 52
  surface, 2, 17, 24ff, 42, 52, 57ff,
   70

safe theorem, 165ff, 178, 186–8
safety factor, 85
  geometrical, 189
St Geneviève, Paris, 182
St Isaac, St Petersburg, 185–6
St Peter's, Rome, 76, 97, 175, 178
sandstone, crushing strength of, 122
sap, circulation of, 198
section modulus, 90–1, 101ff
shear stress, plane of maximum, 72
shear stresses ignored, 71
slip-circle analysis, 156
slip lines, 149–50
slope-deflexion equations, 109
soil tests, 131–2
soil thrust, vii, 2, 15ff, 42, 51ff, 85, 124ff
spheres, small polished, 75–6, 168, 177
square–cube law, 90, 189
stability
  of arches, 98, 163
  of slopes, 159
static electrical charges, 191
strain, maximum elastic, 101, 104
strength of beams, relative and absolute, 92, 95

stress–strain relation, 91, 94, 96–7, 100ff
stress tensor, 112–13
strut tests, 80

three moments, equation of, 109
thrust line
  for arches, 164, 174, 181, 186, 188
  for domes, 178
torsion balance, 191
tower, maximum height of, 14, 50, 81
trench, greatest depth of, 21, 55, 139,
   140ff, 159
Tresca criterion, 118

velocity field, 116, 137
Versailles, 191
virtual work, 91, 135, 175
vis viva potentialis laminae elasticae, 106
voussoirs,
  crushing of edges, 38, 68, 165, 189
  sliding prevented, 38, 68, 82, 162–3,
   170
voûte (en berceau, etc.), 29, 61, 73, 184

water pipes, 91
water supply, 191
wedge of soil, 85, 112, 126, 132ff
workman, useful to, 4, 24, 43, 57, 124

yield stress, large in compression, 71

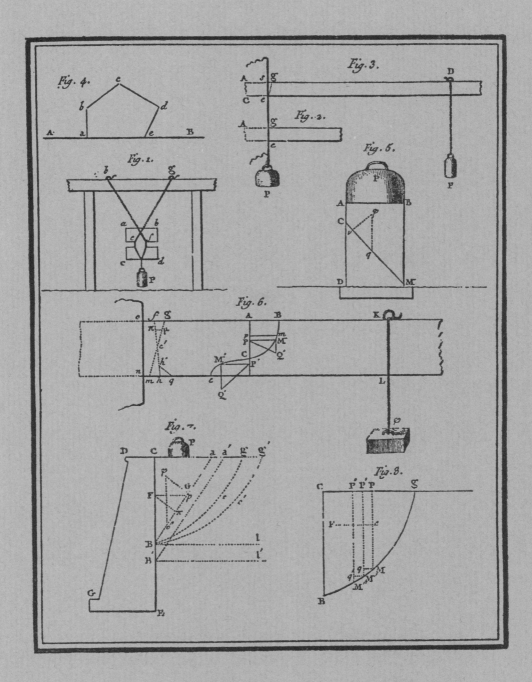